FOUNDRYWORK FOR THE AMATEUR

FOUNDRYWORK FOR THE AMATEUR

by

B. Terry Aspin

Illustrated by the author

ARGUS BOOKS LIMITED

Argus Books Limited
1 Golden Square
London W1R 3AB
England

Fifth Revised Edition 1984
Reprinted 1985

ISBN 0 85242 842 1

Phototypesetting by Performance Typesetting, Milton Keynes

Printed and bound by A. Wheaton & Co. Ltd., Exeter

Contents

30cc lawnmower engine from amateur castings in iron, aluminium and bronze. Total weight of material some twelve pounds.

CHAPTER 1

Introduction

There is scope for the amateur foundryman in the home workshop of any size. Those who possess lathes as small as $1\frac{5}{8}''$ capacity and whose workshop is nothing more elaborate than a quiet corner of the kitchen will, nevertheless, find a useful ally in a No. 1 crucible, which for the melting of aluminium can be heated in the open fire. In such cases it will, perhaps, be found convenient to keep a small box of sand outside, where the moulding operations can be carried out minus any attendant risk of domestic friction, the moulds being carried inside and placed upon the hearth for pouring. It will soon be appreciated that the field of possibilities of the most humble workshop is expanded almost beyond measure by this means; although the process is essentially so simple and primitive, it is none the less effective and the beginner is soon asking himself why he has not been enjoying the benefit of it for years. Although limited to aluminium casting by the restricted melting facilities, such material does, after all, lend itself admirably to the type of tool to be used in the subsequent machining.

Fig. 1 will serve to illustrate the extent of the equipment required for an initial acquaintance with the small scale foundry. With the exception of the crucible, A, which must be purchased and, of course, the moulding sand, B (normally a natural commodity), the remaining items can, quite simply, be made at home. C are two forms of fettling trowels, the one above being intended, primarily, for cutting the ingates. The lower one has many applications of a general nature including smoothing up the joint faces of the moulds and pressing back loose particles of sand. D is the stick for forming the runner. For small aluminium castings it should be, roughly, an inch in diameter. Six inches of broken broom handle is admirable. The pair of moulding boxes or "flask" at E come in for a full description later as do F, the shallow ladle for skimming the dross from the molten metal and G, the crucible tongs.

The more fortunate owner of larger machine tools, with an outdoor workshop and other blessings, will be able to go further and equip himself with a fully fledged foundry on a small scale. He will be able to tackle castings in a variety of metals, including iron, the weight and form of these being governed only by the limit of his own ideas and requirements. Members of societies may like to pool resources and add foundry facilities to the kind of communal workshop more usually encountered. Here equipment could run, perhaps, to a small cupola for melting iron, together with a crucible furnace for the non-ferrous alloys. The interchange of patterns between members would be a boon and the division of the costs between them would make possible the purchase of orthodox metal moulding boxes and other items of useful equipment to the ultimate advantage of all concerned.

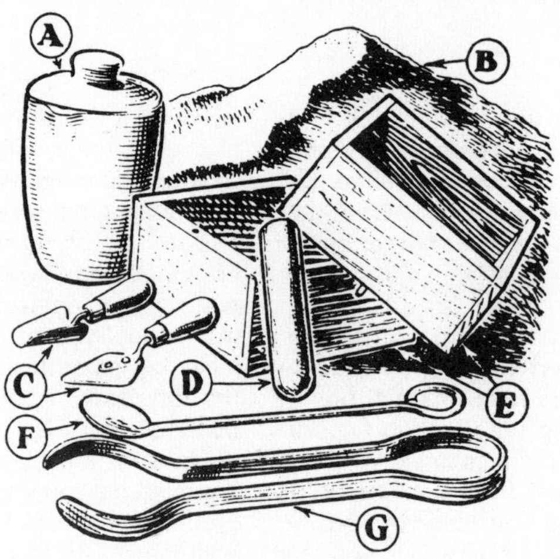

Fig. 1. The simple apparatus required for making a start.

Apart from the obvious advantages to be gained from a venture into the science of the foundry, the added interest involved is a very strong factor on its own. Of the other processes in the workshop, the majority can be directly controlled, by the operator, through every stage to completion, progress being observed, assisted and modified to ultimate success. The result to be expected when a mould is opened after pouring, however, is always a matter for conjecture. A perfect casting cannot be guaranteed every time even in the best of circumstances and although the measure of certainty can be improved to a very great extent by the exercise of care and the gaining of experience, there is always a keen sense of excitement at the moment when the mould is shaken out and the still hot casting emerges.

True, one arrives at the stage when simple patterns are moulded and cast without much thought and a "waster" comes very much as a surprise. There always remains, however, the occasional ticklish job to be encountered which tests the ingenuity of the amateur to the utmost. Thus his interest is stimulated anew and, as with other branches of this most absorbing of hobbies, the deeper he becomes involved and the better his understanding of the process, so also the degree of his enjoyment is increased accordingly.

Foundrywork is put forward as an alternative to the other methods commonly applied, in the home workshop, to the fashioning of an object in metal. Fabricating from separate pieces or hewing from a solid lump of material can both be tasks of extreme tedium. On the other hand, once a little foundry has become well and conveniently established, even a "one off" job can be executed quite expeditiously, certainly with equal if not more satisfaction, by the simple preparation of the pattern in wood, followed by moulding and casting in patterns, moulds, casts and machines and, knowing his own limitations, and those imposed by the capacity of his workshop, fits the job accordingly. The liaison is quite complete. He would be a fool indeed who would provide himself with drawings and patterns, for example, of an object just too large to be accommodated on his own machines, or which demanded a process to complete quite beyond the resources of his own workshop.

Fig. 2. A typical mould, with its pattern.

metal. The pattern remains to be used again if required and soon the foundryman finds himself in possession of a fine stock of useful patterns to fall back upon and he comes to regard them as some of his most valued assets.

Another point is that the amateur foundry offers scope to the hobbyist who would wish to work to his own designs. He prepares his drawings, makes his

The basic requirements of a foundry are simple. First a pattern, a close representation of the object to be cast, in wood. Wood is by no means the only material used for patterns, of course, but for the purpose of the type of work to be discussed here it is probably the most generally favoured. By means to be outlined later, this is embedded in sand contained in a moulding box and later withdrawn to form a cavity, which is again a representation, but in the negative, of the object to be cast. Metal, contained in a suitable vessel, is melted in a furnace or in an open fire and is poured into the mould to fill the cavity. It remains long enough the solidify, when the mould is broken up and the casting extracted.

A typical mould, with its pattern, for producing a most useful pulley wheel, is shown in Fig. 2. This, then, is the essence. It is a process which can be expanded or contracted to fit individual requirements and, although modified or improved according to circumstances, from the most primitive to (in industry) the most highly mechanised and scientific, the same guiding principles apply throughout.

CHAPTER 2

Crucibles and their Care

It is suggested that the reader wishing to experiment with foundrywork on his own account will find that the best results are obtained from the use of a regular crucible or melting pot. There *are* alternatives, of course. An iron saucepan or even an empty can have been known to be used for aluminium. Indeed, the former may have an actual value if steps are taken to protect the interior – or rather, to protect the molten metal against the interior of the pan – by the use of a refractory wash (see appendix). Many metals are soluble in molten aluminium, iron particularly so and its effect is not good, so that bringing the melted metal into direct contact with that of the pan is not recommended. Although iron pots are actually used under certain circumstances in industry, the great disadvantage is that the internal coating is somewhat subject to erosion and, if adequate protection is to be maintained, requires fairly frequent renewal.

Another disadvantage to the use of an iron pot, and this time of particular importance to the smaller foundry, is the rapid cooling, as compared with the crucible, when the charge is removed from the furnace prior to pouring.

In view of the fact, then, that satisfactory crucibles are readily available in a wide range of useful sizes – certainly to suit the amateur – the beginner will find it to his advantage to obtain a small supply at the outset and avoid the sometimes discouraging effect of trying to make do with a makeshift. The type of crucible most commonly used in the foundry is that of fireclay mixed with plumbago (another name for graphite or "blacklead") and, except where special furnace arrangements demand special shapes, the same kind will be found convenient for any of the metals likely to be encountered. Another type, the carborundum crucible, has some advantages in the melting of non-ferrous alloys claimed for it by manufacturers and is worthy of consideration in that field. The common plumbago variety will give equal satisfaction for all metals, including iron.

The shape usually encountered and the sizes most likely to find favour in the small foundry are illustrated, with dimensions, in Fig. 3. The sketch may serve as a useful guide when ordering the first batch. This shows the standard sizes and capacities of the SALAMANDER crucible and is described as the "A' Shape. A.1, although not the smallest size, holds about $2\frac{1}{2}$ lbs.

of molten brass and may be convenient for beginners. At the other end of the scale A.5 can be used for up to 15 lbs. of cast iron and, when at white heat, can be quite difficult to handle by one person using amateur type tackle. The same size, of course, would handle quite comfortably when melting aluminium. All crucibles containing molten metal should be treated with great care and respect and larger pots than this may call for more specialised equipment and, probably, an assistant.

Where different metals are being melted it is generally advisable to retain a separate pot for each one. For the melting of non-ferrous alloys the life of one of these plumbago pots, barring accidents, of course, is almost unlimited. Always care is needed in their handling in and out of the furnace. The top is the weakest part and should not be subjected to unnecessary pressure from the tongs which, in any case, to minimise the risk of the crucible slipping through and spilling the contents disastrously, should always grip about one third of the way down. The practice of levering the crucible about in the furnace with a poker should be avoided. It is better to make any adjustments by lifting it slightly with the tongs and applying the poker to the fire.

About the worse risk attending a pot, particularly in the case of aluminium, is in replenishing the initial charge, which has become molten, with cold metal. Often the larger pieces are kept back until a pool has been formed when, the heat conduction now being fairly rapid, they will melt more quickly. The rapid conduction of the heat, however, can be a danger if the added metal is immersed suddenly. The incoming charge expands and the already melted aluminium, losing some of its heat and tending to solidify simultaneously, is unable to make room for it by rising up the pot. It expands its force outwards and a cracked crucible is the result.

Fig. 3. Sketch showing the approximate relative capacities of "Salamander" Crucibles

One remedy is to have the molten metal at well above melting point – not always recommended in the case of aluminium – before adding to it. Another, perhaps wiser, plan is to lower the additional metal very gently, allowing it to heat up gradually before total immersion.

Used for melting iron, the life of a

crucible is not so great. Nevertheless, they have been known to last for upwards of a hundred melts with great care. In the way of most commodities the price of crucibles has rocketed in recent years so it is prudent to be cautious in their use. Even so, apart from accidental or careless damage, the cost of a crucible will always show an economy when compared with the purchase of castings from a foundry.

The effect on the crucible of high furnace temperature combined with forced draught is one of gradual erosion. It becomes thinner and thinner until it is advisable to discard it before it becomes a danger to handle or breaks in the furnace with the loss of a melt. By this time the crucible will have become reduced in capacity in any case, so the caster will be finding himself short of metal for the mould.

It is sometimes advisable to use some form of cover to the crucible and particularly when melting the more susceptible metals, aluminium, brass and so on, in a solid fuel furnace. Plumbago lids to suit

Fig. 4. Enabling a greater bulk of metal to be included in the initial charge.

Below, Plumbago crucibles. The one on the right has reached the end of its life. That on the left is unused.

the various size of pot are available and their cost is relative. Their life is also about the same, so it is convenient to purchase pot and cover together. Their use will go a long way towards protecting the molten metal from some of the effects of the furnace atmosphere and to exclude foreign matter such as ash and cinders.

An alternative to the normal cover and which has the added advantage of permitting a larger initial charge is shown in Fig. 4. In this case a second crucible, which could be a cracked or damaged one, is up-ended over the first with the charge piled up inside it.

The once prevalent belief in the practice of annealing a new pot by leaving it to stand, upside down, overnight in a dying furnace, would appear to be unnecessary. Returning the empty crucible to the furnace after pouring, except, of course, for a fresh melt, would also seem to be an overrated superstition, merely contributing towards a shorter crucible life. It may not be advisable to cool them too suddenly, however, and usually a warm spot can be found somewhere behind the furnace. The "kitchen fire" foundryman may find it convenient to place his pots in the ashpan to cool. In general, it may be said, plumbago crucibles will be found to withstand a remarkable amount of fair use and even rough handling.

CHAPTER 3

The Furnace

Of primary importance in any foundry, large or small, are the arrangements for melting metal. It has been indicated that, where castings are to be limited to a small size (up to a pound or two in weight) in aluminium, no special furnace is required at all. The metal melts at a dull, red heat (1,200°F or thereabouts according to the alloy), which can easily be attained in a kitchen range. Thus, there is little excuse, on grounds of lack of facilities, for failing to avail oneself of the advantages to be gained by a modest acquaintance with foundrywork. It is usually advisable to make use of a bright, clean fire, to avoid the inclusion of soot in the casting and, often, it will be found that the conventional fireside tongs will answer the purpose of handling the pot.

Having progressed beyond the scope of the open range, the amateur foundryman will probably become interested in a furnace of larger capacity. Among the ready-made possibilities in this direction is the ordinary slow combustion stove. While the purchase of a new one for that specific purpose could hardly be recommended on the grounds of economy, there may be the chance of picking up a second-hand stove fairly cheaply. In that case there need be no compunction about making the necessary modification to fit it for its new role.

For making castings machinable on the popular $3\frac{1}{2}''$ lathe, the stove need be no larger than fifteen to eighteen inches high. A bigger stove would only use a greater quantity of fuel than should, otherwise, be necessary and nothing is to be gained by that. It will usually be advisable to take away the cast iron top to facilitate the entry and removal of the crucible and, if the furnace is to be built into a flue to make use of natural draught, some further slight modification will be required (as shown in Fig. 5) to admit the end of the stove pipe. In this case the usual recess in the back of the stove has been cut lower and the lining made good by the use of one of the proprietary fire cements. The same can be put to excellent use in repairing any defects in the lining which may appear from time to time. Neglect of this point will probably result in the wrapper being burned through, as the lining of a slow combustion stove is comparatively thin. A furnace exactly as illustrated has been known to make hundreds of castings in aluminium, brass and iron.

Always, when solid fuel is used, the provision of natural draught is a real advantage in lighting up. Except for this,

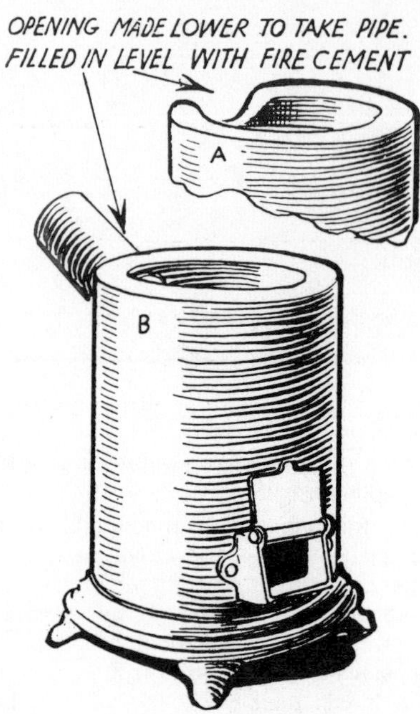

Fig. 5. Slow-combustion stove converted to a furnace.

however, excellent use can be made, out in the open, of the stove alone, provided an adequate forced draught is available. For anyone not wishing to go to the expense of a blower, a very high temperature can be reached with the aid of a fairly tall chimney. Aluminium can be melted with ease and there is a possibility of being able to handle some of the cuprous alloys, too. About eight to ten feet of four inch pipe will be found sufficient.

Although the slow combustion stove has been put forward here as a practical solution to the furnace problem, no doubt there are many other possibilities which will suggest themselves to the amateur. One rather obvious alternative is to build a square section furnace from firebrick. The same is rendered more serviceable if the bricks are enclosed within a sheet-iron wrapper. An example is illustrated in Fig. 6 and, being built to the convenient dimensions of the brick, has a furnace cavity of really generous capacity.

Although in the ordinary way crucible furnaces are equipped with fire bars and ash pan, such refinements have been deliberately omitted from the versions shown. Where the fire is by ordinary gas coke, fire bars could prove a nuisance by becoming clogged with clinker. These furnaces can easily be cleared while the work is in progress if the need arises.

A "Monolithic" Lining

In the industry, nowadays, many furnace linings follow the type known as "monolithic." The term is more descriptive than correct and refers to the kind of lining rammed up on the spot, from suitable refractory materials, within the cavity formed between the furnace wrapper and a suitably constructed temporary former. Fig. 7 is an illustration of a small example of such a furnace making use of a disused five-gallon drum.

The ends are first removed with a chisel and hammer and a hole cut close to the top to admit the stove pipe. Another opening, which serves, collectively, as an ash hole and for blast, is cut at the bottom. A useful former, the diameter of the furnace cavity, can be rolled from sheet metal and braced internally with wooden struts. It should be constructed so that with the wood supports removed the metal will spring inwards to facilitate its removal after the lining has been rammed.

Material used for the lining is fireclay mixed (about fifty-fifty) with "grog" – the term used to describe the finely ground up firebrick with which the clay is reinforced – to improve its strength, its resistance to high temperatures, and to reduce the tendency to cracking. The addition of grog also allows of the lining being dried out more quickly than would otherwise be possible.

Those living in an industrial area will experience little difficulty in obtaining the necessary materials, if an approach is made to the management of one of the local brickworks. The quantity required is small and, if charged at all, the cost is negligible. The quality of the clay will vary but there is little doubt that for the type of furnace in question, results will be quite satisfactory.

In the event of the district being remote

Fig. 6. A furnace built of brick.

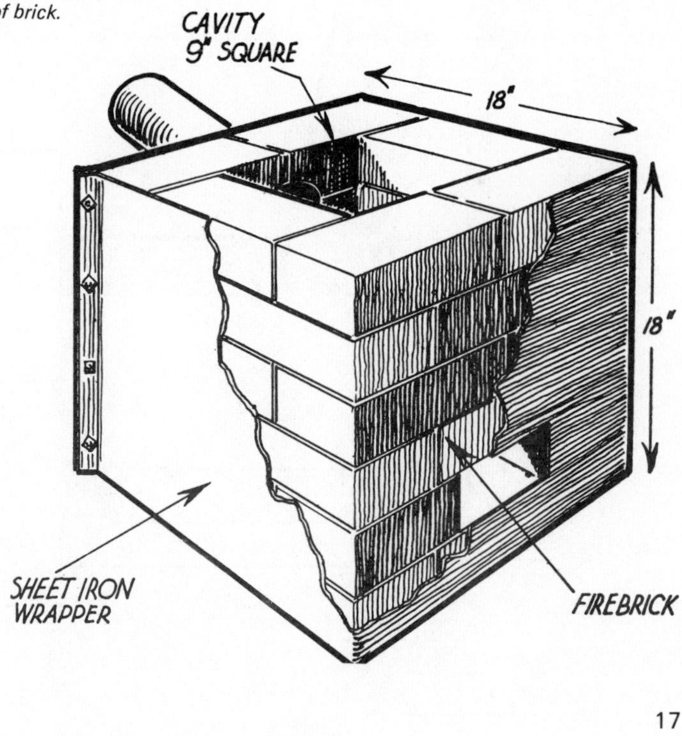

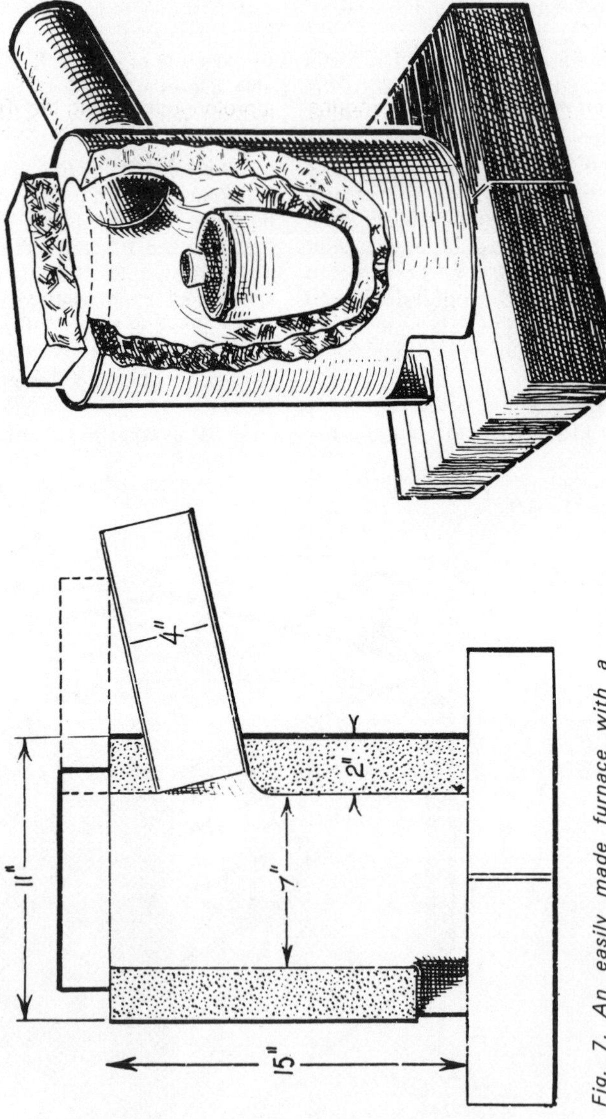

Fig. 7. An easily made furnace with a "monolithic" lining.

from industry, it will be found possible to obtain a superior quality clay from one of the crucible manufacturers.

The clay and grog are received dry and amalgamated as thoroughly as possible by hand. The material is then mixed with from five to ten per cent water to produce a heavy clay to about the same consistency as moulding sand, i.e., bonded but not over-saturated. It is fed into the space between the former and the furnace wrapper and rammed down solidly, the feeding and ramming being carried out in a spiral, round and round the furnace and as continuously as possible. A break in the operation may result in a crack. To obtain a lining of unvaried strength throughout, particular attention should be paid to the ramming to keep it consistent.

Before knocking away the wood supports and removing the former, the end of the flue pipe can be inserted and the vent below opened out and cleaned up with a trowel. The same can be applied to any raggy parts to tidy the lining up generally, before it is left to air dry for a period – not less than twenty-four hours.

Firing the Lining

The drying out can now be accelerated by lighting a small wood fire in the furnace and keeping it going until the clay has ceased to steam. A gradual increase in the temperature can then be permitted by the addition of coal and coke until, after perhaps a couple of hours, it is possible to bring the whole of the interior of the furnace to red heat. At this stage it can rapidly be blown up to working temperature and allowed to remain at that for two or three hours, or more if possible. Finally, it should be packed up with coke and left to burn itself out over night.

If everything has gone according to plan a good, sound lining should result. Tapped lightly with the fingernail it should give the characteristic "ring" and the colour should be a light, yellow brown.

Having gone to a good deal of trouble to make a really serviceable furnace, it will be well worth observing a little care in its use and, thus, enjoy the full advantage of a prolonged working life. The main enemy is, probably, clinker. It usually forms pretty badly in the lower part and there is always a strong temptation to attempt its removal by pounding with the poker. Nevertheless, sometimes it may be preferable to leave a particularly tenacious lump adhering to the lining than to risk it bringing a piece of the same away with it, if forcibly dislodged. Sometimes it is possible to clean out a good deal of the clinker while it is in a semi-solid state in the red hot furnace, immediately after the final melt.

No reference has yet been made to the method of closing the furnace mouth while the melt is in progress. Generally it will be found that the most satisfactory cover is a slab of firebrick, particularly in cases where the higher temperatures are reached. To give access it is lifted off or pushed aside with the tongs. The rather obvious idea of tilting the cover backwards against, perhaps, the foundry wall is liable to roast the operator. When the melting of iron is in progress the cover lifted thus radiates a generous contribution of white heat.

Furnace covers of firebrick, subject as they are to local heating, are very prone to crack. It is a real advantage if they are easily and cheaply replaceable.

Fuel

Each of the furnaces so described is intended for the use of solid fuel. This does not exclude the possibility of gas or oil firing, of course. There is room for experiment in both these fields from the amateur point of view. There are, however, two rather strong advantages to

Fig. 8. Some vacuum cleaners are readily adapated as blowers.

the suggested use of coke. The first is that, in one form or another, it is a commodity which is pretty universally obtainable. Secondly, where the melting of iron is anticipated, it is probably possible to obtain a far greater heat with simple apparatus than with either of the other two. In this respect, it certainly compares well with oil firing, which requires a special burner as well as a means of providing blast.

It is well worth the trouble of anyone taking up foundrywork seriously to make an effort towards obtaining a supply of industrial furnace coke. This material can be recognised by its silvery-grey colour and it burns with great heat without an excess of clinker forming in the furnace.

An excellent alternative can usually be obtained from solid fuel merchants and may come under the title of "Sunbrite" or some other modern designation. The domestic supplier should be able to help with the kind of small coke he delivers for use in domestic boilers. The beginner should not be deterred by any difficulty in obtaining what he believes to be the most suitable fuel for the job. Some effort should first be made using that which is easily available. After all, early founders in iron and bronze used wood charcoal!

Forced Draught

Where blast is necessary, there is little doubt that electric blowers of one form or another provide the most convenient answer. Hand forge blowers or large bellows could, of course, be used as an alternative but, since a forced draught must be sustained over comparatively long periods, it is obvious that the human energy expended will be rather considerable. With the fairly high temperature in the neighbourhood of the furnace the effort is not going to be performed in comfort.

In cases where electricity, mains or otherwise, is available then it will usually be a great advantage to obtain some kind of small blower. There are many offered, by advertisers of surplus equipment, at very reasonable prices for use on low voltage. Choose a large one for preference even if the present furnace is quite small. A surplus of air is more useful than too little.

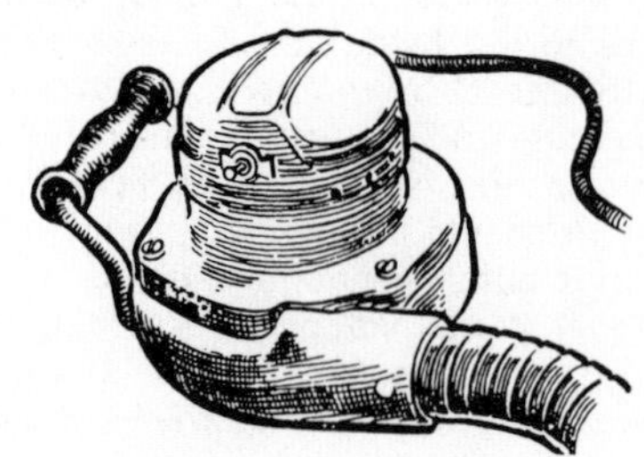

Fig. 9. Ex Air Ministry, 230 volt blower.

Those able to "borrow" a vacuum cleaner from its more normal function will have an adequate supply of air readily to hand. Fig. 8 shows the adaptation of a popular type of vacuum cleaner which is provided with an arrangement for utilising the exhaust as a standard fitment. A machine which is admirably suited to the job is the low pressure type of paint spraying blower and, no doubt, there are many workshops already equipped with such an appliance. Fig. 9 is an example of a 230 volt blower purchased from Air Ministry surplus. The motor is rated at one kilowatt, and with a two inch delivery pipe of aluminium flexible tubing, it is able to supply a very comfortable draught to any of the furnaces discussed. The simple method of directing the blast into the bottom opening of the furnace is illustrated in Fig. 10. A rough, sheet iron cowl serves to prevent cinders and dust blowing out the wrong way.

Another kind of blower suitable for furnace work is the Root sliding vane type, Fig. 11. Generally speaking it is the volume of air delivered which counts for foundrywork. Pressure per square inch need be no more than five pounds.

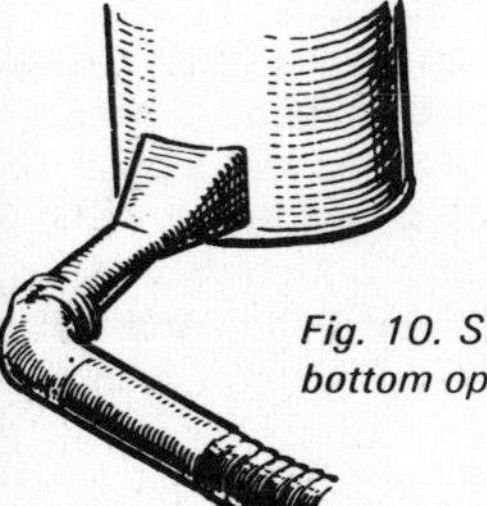

Fig. 10. Showing how the blast is led into the bottom opening of the furnace.

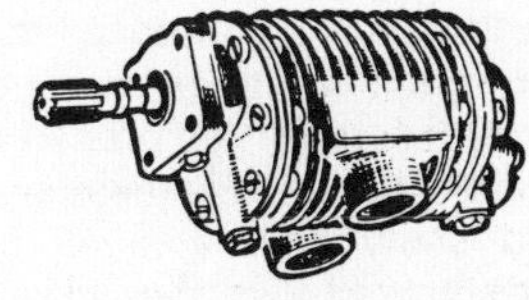

Fig. 11. Sliding vane blower.

CHAPTER 4

Sand

It could be that, from an amateur point of view, the greatest obstacle is the acquisition of a suitable sand, for, although it is a commodity which forms one of the most abundant natural deposits on earth, by no means all of it has the inherent properties which fit it for the preparation of a mould. Nevertheless, it is the purpose of this book to show how a small foundry can be assembled with the aid of materials most readily to hand and, in this respect, sand is a particular case in point.

Perhaps a choice of sand will be assisted by a better understanding of the function required of it. To begin with a relatively fine grain size is desirable to reproduce the detail of the pattern, to reduce metal penetration and to impart a smooth and pleasing appearance to the casting. The sand must be of a refractory nature to withstand the high temperature where it comes into contact with the molten metal and, in a mass, it must be permeable to allow the free passage of gases generated in the mould during casting. Another important feature is the property of the sand to hold the form given to it by the pattern. This is known as "bond."

Except, perhaps, where permeability is reduced by an over-abundance of clay, the first three of these conditions are usually encountered automatically in natural sands. There is, of course, a wide variety in grain size. The matter of bond, however, is a property which suits one deposit of sand to foundrywork rather than another. Given a sand which, though lacking in bond, fulfils the other conditions, this latter property can be adjusted artificially in many ways.

Many amateur foundrymen get over the difficulty of obtaining suitable sand by making a friendly acquaintance with someone in the industry, where a hundredweight or so is never missed. Others will be fortunate in living in an area where acknowledged deposits of foundry sand occur. For the purpose of the home foundry this can be used exactly as supplied from the sandpit. Although additions of coaldust or even horse manure may be recommended by the Old Hand there is no real necessity for either, particularly the latter, in the handling of the relatively small castings involved. A quantity of Mansfield sand, which is used extensively in brass founding, would probably represent a particularly valued asset to the small scale foundry.

For others, who may not be so conveniently placed, there is a case for using sand of any kind from possible local deposits and adding to the quality of its bond. Red building sand is not without its possibilities and this may be strengthened by the addition of natural clay or even fireclay, while fuller's earth, obtainable from any chemist, can also be put to good effect. The latter, under the name of *Fulbond*, is supplied in various grades particularly suited to foundry requirements and is also valuable as a regenerating agent for used sand which has become lacking in bond. Sea-shore sands, usually referred to as silica sands, can be – and are – used for the purpose of mould making. Industrially, sea-shore sands are used extensively in the making of sand cores: largely because of their permeability and because they possess little or no natural bond. The latter fact may appear as a paradox where the matter of bond is of such prime importance but, in that case the bonding agent, usually organic, burns away on casting and the core empties freely when the mould is shaken out.

Sea-shore sand most suited to moulding purposes is usually of the "wind blown" variety which forms from the sandhills round our coasts. For the best results it should be free from particles of shell, which are not so refractory as the sand itself. Again, in grain size, sea-shore sand varies considerably between one area of coast and another. One is as fine, almost, as pepper and another is much too coarse to be used at all. Among the localities recognised as sources of supply for example, are Skegness, Southport and Prestatyn, but there are many more. For ordinary green sand moulding experiments may be made using silica sand mixed at the rate of about one in twenty with fuller's earth. While it is not pretended that the moulding qualities will be ideal, it will hold the impression very well and, for aluminium, where a little more moisture may be tolerated in the sand, it will give the beginner every chance of success.

Moisture Content

Before use for mould making the sand must be "tempered" with water. In fact, apart from moistening the bonding agent, the surface tension of the water forms part of the green bond. The quantity of moisture is in the region of two pints of water to the hundredweight of sand but, as it is unlikely that the amateur will wish to weigh out his sand each time he commences work, it will probably be more convenient to judge the tempered sand by its characteristics. In time, the benefit of his own experience in the matter will serve as the best guide but, as an indica-

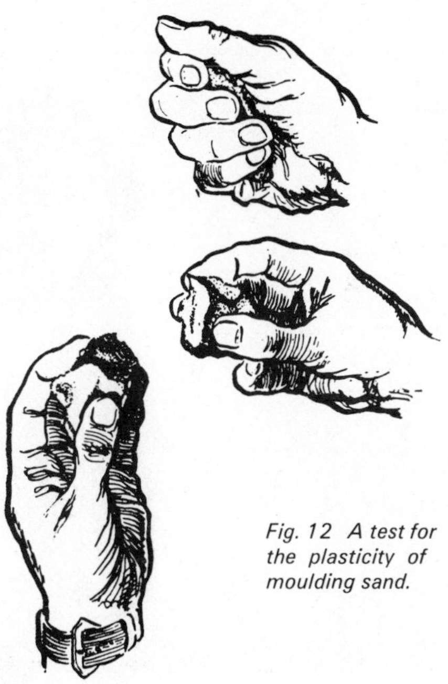

Fig. 12 A test for the plasticity of moulding sand.

tion, one test is to squeeze a handful in the palm and then break the lump in two, Fig. 12. The fracture should be clean, without "dribbling," showing adequate moisture.

If an attempt is made to use the sand too dry it will, probably, result in parts of the mould falling away, often at the moment when the box is being closed again after the removal of the pattern, commonly known as a "drop-out." Very annoying if a good deal of time and care has just been expended in preparing the mould, as often is the case when a drop-out occurs. Sand which is too wet will usually reveal itself when the mould erupts like a miniature volcano as soon as the metal enters. Such a condition can, of course, be dangerous.

Methods of tempering the moulding sand will, of course, vary but one of great convenience is to spread the sand about an inch thick over an area of floor and, after sprinkling with water, mix thoroughly with a garden rake. It is usually quicker and more effective than using a shovel. The sand should be passed through a quarter-inch mesh sieve after mixing and it will be an added advantage if it is left to stand for half an hour or so before moulding commences.

When the mould is prepared thus and used in its damp state to receive the metal the process is known as green sand moulding and the strength of the sand is known as its green bond. Sometimes, however, a mould is dried out, either fully, in an oven, or partially (skin dried) by the

Fig. 13. Tempering moulding sand with rake and sieve.

Wheel castings for a 5" gauge locomotive just as they left the sand. Note the oval spoke section.

application of a flame from a gas torch or other. Another form of skin drying is the "shell" mould, an industrial application, where the mould is rammed up on a machine using a heated metal "plate" pattern. The sand coming into contact with the pattern is thus given a hard skin or shell, with added strength for resisting the erosive action of the incoming metal.

Oil Sand

Where a mould is intended for drying out by baking it is usual to make it from oil sand. Sometimes the oil sand is used only for facing, as it is comparatively expensive, and in that case it is packed round the pattern and the rest of the box is filled in with green sand. Oil sand is used, also, for coremaking, particularly in the case of the type of core used in the production of light machine parts and model castings. At the same time it provides a particular problem for the amateur because oil sand cannot satisfactorily be mixed by hand.

Usually the basis is silica sand to which is added a proportion of a synthetic bonding agent and water. In actual fact, although the designation is "oil sand" the proportion of oil actually used in the mix is usually only about half that of the moisture content and smaller, also, than the proportion of any additional bonding agents which may be employed. Having no natural adhesion, the silica sand must be reinforced with a material to form its green bond and, after baking, its dry bond. In the former case the addition may be molasses, dextrin or a proprietary core gum or cereal binder and, in the latter, a drying oil, such as boiled linseed oil, is used. Sometimes a proportion of red moulding sand is used with the silica sand with a view to varying the condition and strength, in the green or the dry state as the case may be, and the amateur will also find room for experiment in this respect.

A typical formula for oil sand would be in the nature of one hundredweight dry silica sand, two pints water, two pounds dextrin and one pint linseed oil. Of course, the quantity required at any one time in a home foundry would be very much less than this and, in all probability, if fourteen pounds of sand were used, with the remainder of the ingredients pro rata, the bulk would be ample. The mixed sand will keep for a considerable time in a closed and air-tight tin and the slight mould which forms on the crust is of no detriment.

While the remaining ingredients are common enough, dextrin is one which may be something of a problem to procure. It appears disguised by one name or another as a product of a number of firms specialising in foundry supplies and, in its usual form is in the nature of a strong gum (British gum). A very convenient form of the material and one which is fairly readily obtainable is marketed under the name *Dextrine,* and can often be purchased, along with the linseed oil, from the usual paint stores or oil shop. With it, a very useful form of oil sand can be prepared as follows.

An eighth of a pint ($2\frac{1}{2}$ fluid ozs.) of linseed oil is added to fourteen pounds of sea-shore sand tipped on the floor and the two combined, as thoroughly as possible, with the aid of a trowel or small boiler shovel. The oil will be found to mix in readily enough. Half a pound of *Dextrine* is made up, in accordance with the instructions on the packet, with a quarter pint of water and about half the resultant quantity of gum is added to the sand and mixed in. At this stage the result will be anything but satisfactory, being composed of clots of gluey sand and much loose stuff very difficult to amalgamate. The amateur will begin to appreciate the value and importance of the mechanical methods of mixing available to the professional foundryman.

If, however, he possesses an electric hand drill of the *Wolf, Black and Decker* or similar type, he has an admirable substitute close at hand. About half the above quantity of sand is conveniently dealt with at a time, where it is contained in a small bucket or empty paint tin. Fig. 14. The simple beater shown is bent from a length of quarter-inch diameter mild steel rod and the revolving drill is passed round and round inside the tin when, in perhaps two or three minutes, the sand will be quite thoroughly mixed. The second lot can be treated in the same way and, if it is then found that the mixure is a little dry, more of the *Dextrine* gum can be added and the process repeated until the correct consistency is obtained. In any case, repetition is more valuable than otherwise. The sand cannot really be over-mixed.

To explain the matter of the correct consistency of oil sand is a problem. Anyone who has had access to a quantity of it from a foundry will have some idea what to expect but, for others, it will probably be best to compare its characteristics with those of common brown sugar, except of course that it is not sticky. It feels soft and smooth and, when the hand is pressed into it, it holds the form. If it sticks to the hand at all either ineffective mixing or an overabundance of moisture or gum is indicated. When finally turned out of the can after mixing it should stand up in a soft, but firm, cake.

Sand so prepared may be somewhat lacking in green strength. Improvement in this respect can often be effected by the use of upwards of fifty per cent of red moulding sand with the silica sand or, alternatively, by a small addition of fuller's earth. The foundryman can often make adjustments in this way until he has a material entirely to his liking. On the other hand the dry strength after baking may be found to be greater, in the case of the mixture made up entirely from silica sand, than it is in the modified version.

Moulds and cores made from oil sand may be baked, quite effectively, in an ordinary domestic oven. For best results the temperature should be round about the 400°F mark – say 200°C – the temperature required for cooking meat. The baked colour will be somewhat darker than the sand in its green state and the surface of the core should be quite hard when tested with the fingernail.

Where facilities for making oil sand at home are not available to the amateur, or when a particularly exacting core or mould is being attempted, it may be an advantage to obtain a quantity ready mixed, either from a local foundry or from a supplier. At least one firm (see appendix) will supply a high quality sand in quantities as small as a quarter-hundredweight. It is ready for use and will keep for a long period in an airtight receptacle, and, when twenty-eight pounds will make many, many cores, it cannot be regarded as a great expense. The method of drying out is the same as that employed with home mixed sands and a temperature of 400°F is equally correct. The sand is baked until it is an even, chocolate brown in colour.

Loam

Occasionally the amateur foundryman will be confronted with a reference to loam (colloquially "loom") moulding. It is a technique outside the scope of the home foundry and usually describes the material used in the preparation of moulds and cores associated with "pit" moulding, which is carried out on the floor of the shop and deals with castings to upwards of many tons. Loam is a synthetic moulding material composed of sand which has been finely milled, and an addition of chopped straw, horse manure or wood sawdust. Occasionally, horsehair is added to improve its tenacity. The work is highly skilled and moulding is largely carried out, by hand, from templates or "strickles," as they are called, rather than patterns in the usual way.

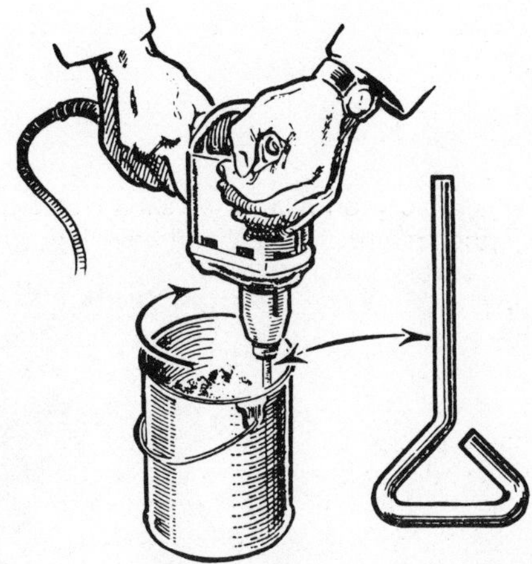

Fig. 14. The use of a portable electric drill for the mixing or "milling" of oil sand.

Blacking

The use of powdered graphite or "blacking," also called plumbago, applied to the surface of the impression in the mould has the effect of assisting the flow of metal. It also imparts a higher finish to the casting than would occur if the metal were brought into intimate contact with the sand. The tendency of the latter to adhere to the surface of the casting is also

The 30cc lawnmower engine shown earlier at a more advanced stage.

much reduced. Graphite is normally used as a mould dressing for iron and cuprous alloys. Aluminium is frequently poured into uncoated moulds.

Green sand moulds are dusted with dry plumbago, which is normally contained in a little cotton bag for the purpose. The bag is shaken over the mould allowing the powder to fall and penetrate the cavity. A useful alternative is a fine, kitchen sieve or strainer, which is tapped lightly with the finger to release the graphite. When powdered plumbago is used as a dressing for cores and dry sand moulds, it is usual to mix it with water and apply it with a soft brush. The core or mould is then dried off again before it receives the metal.

Powdered graphite is obtainable from

crucible manufacturers. Another form of refractory mould coating is composed of graphite or other material suspended in a liquid, usually a spirit, base. This is most conveniently applied by spraying and is suitable for application to either green sand or dry-sand moulds and cores. Grades suitable for ferrous and non-ferrous metals and light alloys are available and they have the added advantage of being inflammable. They are ignited after application and impart a skin-dried effect to the mould.

Fig. 15. The use of a spray gun for the application of refractory dressings to the mould.

Parting Sand

Sand which has been burned loses its bond. Thus a sprinkling of burned sand is applied to the joint face of the mould, before ramming up the cope, so that the two halves will come cleanly apart. Dry sea-shore sand may also be used for the same purpose while there is, occasionally, a case for separating a mould by means of strips of newspaper.

Special parting powders, which perform most efficiently, and which may be composed of bone dust or produced from gypsum, are available and it is well worth the trouble to procure one of these as work progresses. Sometimes they can be of particular value, when dusted over a pattern, to assist in a clean withdrawal. In the making of oil sand moulds and cores *liquid* parting, applied by brush or spray to the pattern or corebox, is exceptionally useful in affording a clean strip. This again, a water repellent liquid, can be obtained from firms of foundry suppliers. When spraying liquid parting or applying liquid dressings to the mould in the same way, it is advisable to wear some form of respirator. If the only actual ill effect, the nasal irritation can be very bad!

Portland Cement

Perhaps, before leaving the chapter on moulding sands completely, a brief reference to one further technique will not be out of place. It may be of interest to some, not very conveniently placed with regard to the more orthodox sands, to learn that ordinary *Portland Cement* can be and is used as a bonding agent. Known as the *Randupson* proces, it concerns the use of silica sand mixed with ten per cent cement and, approximately, five per cent water. Moulds should be air dried for twenty-four hours and may then be dried out more rapidly.

CHAPTER 5

Moulding Boxes

For the kind of work likely to be encountered in the small scale foundry, the sand is rammed into a pair of moulding boxes to form the mould. The pair of boxes are known as the "flask," which is composed of a lower box (the "drag") and an upper box (the "cope"). In foundries, flasks can be of cast iron or fabricated from sheet metal. Nevertheless, an excellent substitute for the metal box can be made from wood and contrary, perhaps, to first impressions, there is virtually no risk of scorching, even when iron is poured to within half an inch of the sides of the box. The moulding sand is such a good insulator. Being much lighter in weight than the metal flask, special attention must be paid to weighting the top of the mould when anything like a substantial section of casting is being poured. Otherwise the cope will probably lift, with a loss of metal, damage to the box and the waste of a mould. Perhaps it is useful to acquire a habit of weighting the boxes

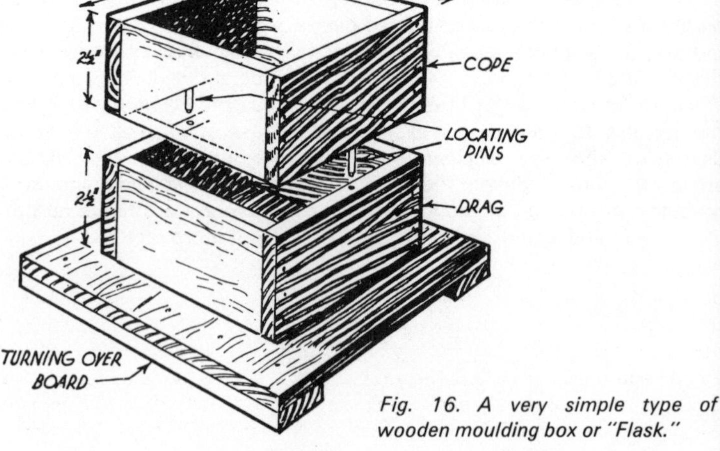

Fig. 16. A very simple type of wooden moulding box or "Flask."

every time. At least *one* possible source of failure is thus eliminated.

A suitable form of wooden flask is illustrated in Fig. 16. This is of a most elementary nature and simply nailed or screwed together from rough sawn timber. Even if an improved appearance is imparted to the outside by planing, it is still desirable to leave the inside rough as a key to the sand. Where planed wood is used it will be advisable to gouge grooves on the inside as an alternative, Fig. 17.

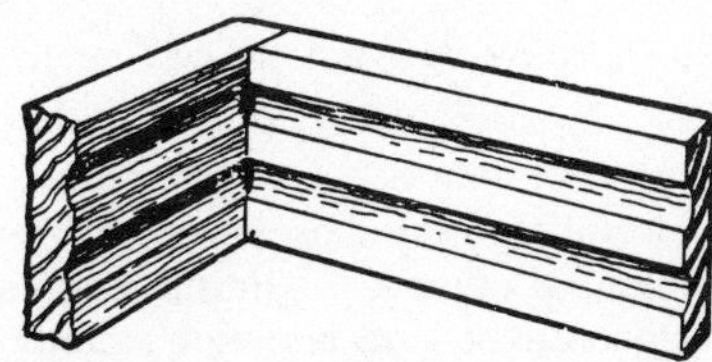

Fig. 17. Grooving the inside of the box.

The small, battened board underneath is a handy addition and comes in useful for inverting the mould, when that is necessary, and for carrying the mould about.

Fig. 18. Locating pin.

Fig. 18 shows the arrangement of dowelling. A beheaded nail is driven into the edge of the cope and a corresponding hole bored in the drag to receive it. It is an advantage if this latter hole is bored clean through the wood to allow any sand, which might otherwise clog it, to drain straight through. The purpose of the dowels, of course, is to afford exact location to the two box parts, ensuring that the two halves of the mould register accurately. This method of dowelling is rendered more exact and less subject to wear if the drag is reinforced with a small metal plate. Fig. 19.

Fig. 19. Metal plate to reduce wear.

A more elaborate form of wooden moulding box is shown in Fig. 20. Obviously, as the size of the box is increased, so also must be its strength to carry the additional weight of sand. A box ten inches square may be close to holding a quarter of a hundredweight. Similarly, while it is a simple matter to close one of the smaller boxes by first registering one dowel and, by lowering the cope by degrees, bringing the other dowel into line with its socket, when a similar technique is employed with a heavier box, the supporting of the added weight can be fatiguing, just at a time when there is little margin for error. The type of dowel shown here keeps the pins within the range of vision of the moulder, while he stands over the box and closes its more comfortably, Fig. 21. Once the ends of the pins are engaged the register is effected automatically as the cope is lowered.

The brackets holding the pins are cut from a section of angle iron and secured to the box side with screws. A convenient size of pin, which will probably clear the most irregular parting line likely to be

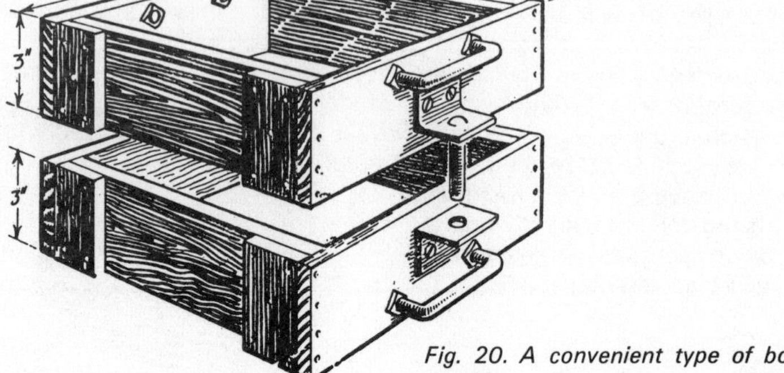

Fig. 20. A convenient type of box for larger moulds.

encountered, would be about two inches long and turned from a piece of $\frac{3}{8}''$ diameter steel. A quarter inch hole is drilled in the cope bracket and the $\frac{3}{8}''$ steel is turned down at the end to fit it, with a little additional length to the spigot for riveting. Location will be assisted by giving a substantial chamfer to the other end which, of course, fits into a $\frac{3}{8}''$ hole in the drag bracket, and is given a comfortable clearance, Fig. 22.

Lifting handles can be provided to advantage. A simple type of handle is illustrated and takes the form of a staple,

Fig. 21. Closing a large mould.

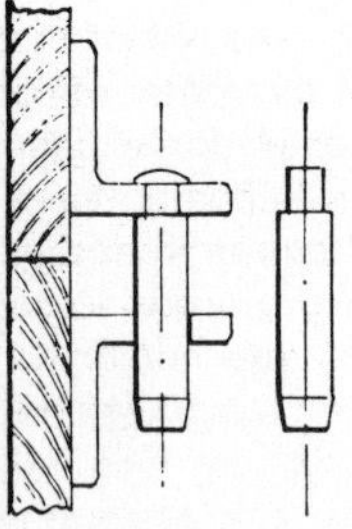

Fig. 22. A pin with angle iron brackets.

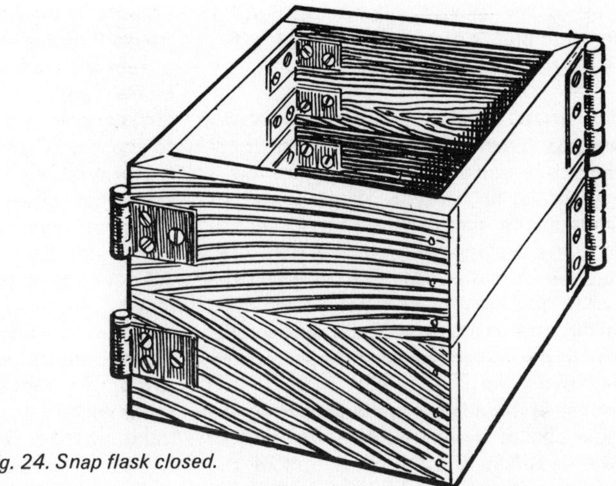

Fig. 24. Snap flask closed.

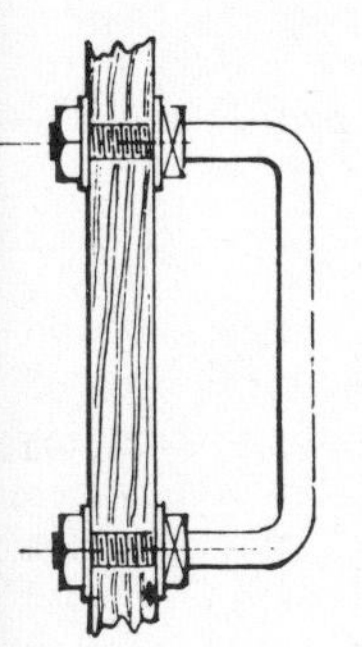

Fig. 23. Bow handle.

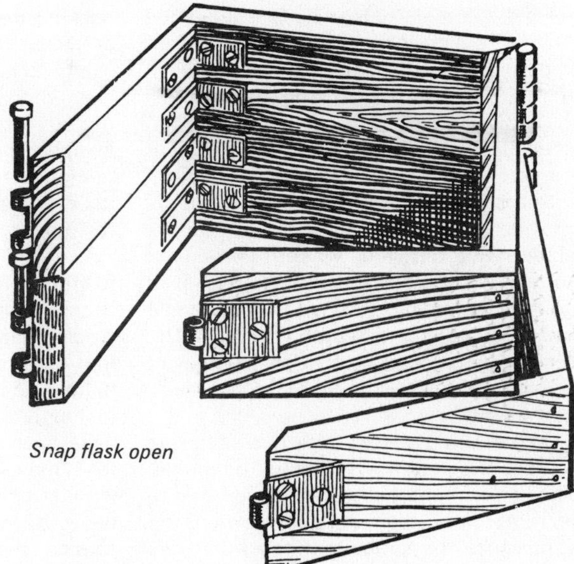

Snap flask open

bent from quarter inch diameter steel rod and screwed at each end for nutting to the sides of the box. A useful location for these handles is directly over the pins, Fig. 23.

The "Snap" Flask

Gaining in favour in mechanised foundries, and capable of adaptation on a small scale, is the type of moulding box known as the "snap" flask. In this case the box parts are made in such a way that, after the mould is assembled and closed prior to pouring, the flask can be removed leaving the sand mould standing on its own. In that way the moulding box can be used over and over again, an unlimited number of moulds being made from only a single pair of box parts. The economy of boxes is offset to a certain extent by the extra trouble entailed in making them; in fact, the cost of such flasks when metal ones are purchased may be four times the price of the ordinary type, but the system is particularly advantageous when applied to moulds which are to be dried or baked.

Shown in Fig. 24 is a form of wooden snap flask which can easily be made at home. The box is hinged at opposite corners. One hinge of each part is provided with a removable pin so that the sides swing outwards when it is required to remove it from the mould. The fixed corners are reinforced with metal angle straps or a short length of iron. The register pins could be similar to either of the examples shown, according to the size of the box.

When using a snap flask it is advisable to keep the pattern well within the mould. The half-inch margin previously referred to is rather less than adequate when the sand lacks the support of the box sides and the risk of break through is increased. Normally about double the usual amount of margin is allowed and obviously conditions will depend upon the weight of the section of the casting and the metal to be poured. It is rather a matter for experiment. Perhaps the most obvious application of a snap flask is in the preparation of baked sand moulds, where the box parts are removed and the sand alone conveyed to the drying oven. Where the halves of the mould are required to be handled separately after drying, for the assembly of cores and so on, hand holds are provided in the moulds themselves as also are locating dowels inserted in the joint faces.

When making any kind of moulding box, it is worth while obtaining a flush and smooth fit between the two halves. The sand will thus be rammed in solidly all round and the risk of stray particles falling into the mould is much reduced.

Flasks for brass founding traditionally take a different form. In that case two backing boards are provided and clamps are used to hold the parts together. An example is illustrated in Fig. 25, which also shows how the metal is poured through holes in one end of the box. The extended pouring gate serves as a trap for foreign matter entering the mould and a good "head" is provided by the length of the box.

Metal Moulding Boxes

In the ordinary way the amateur engineer would never hope to make for himself all the equipment needed to carry on his hobby. So it is, also, with the foundryman although, so far in this book, only crucibles have been referred to as being a "must" for purchasing. An enormous amount of work can be done with the kind of wooden boxes described, but there is always the chance that, progressing to the stage where, from the point of view of interest, foundrywork becomes something of an end in itself, the amateur may feel the

Fig 25. Brass founder's flask, and, below, pouring through one end.

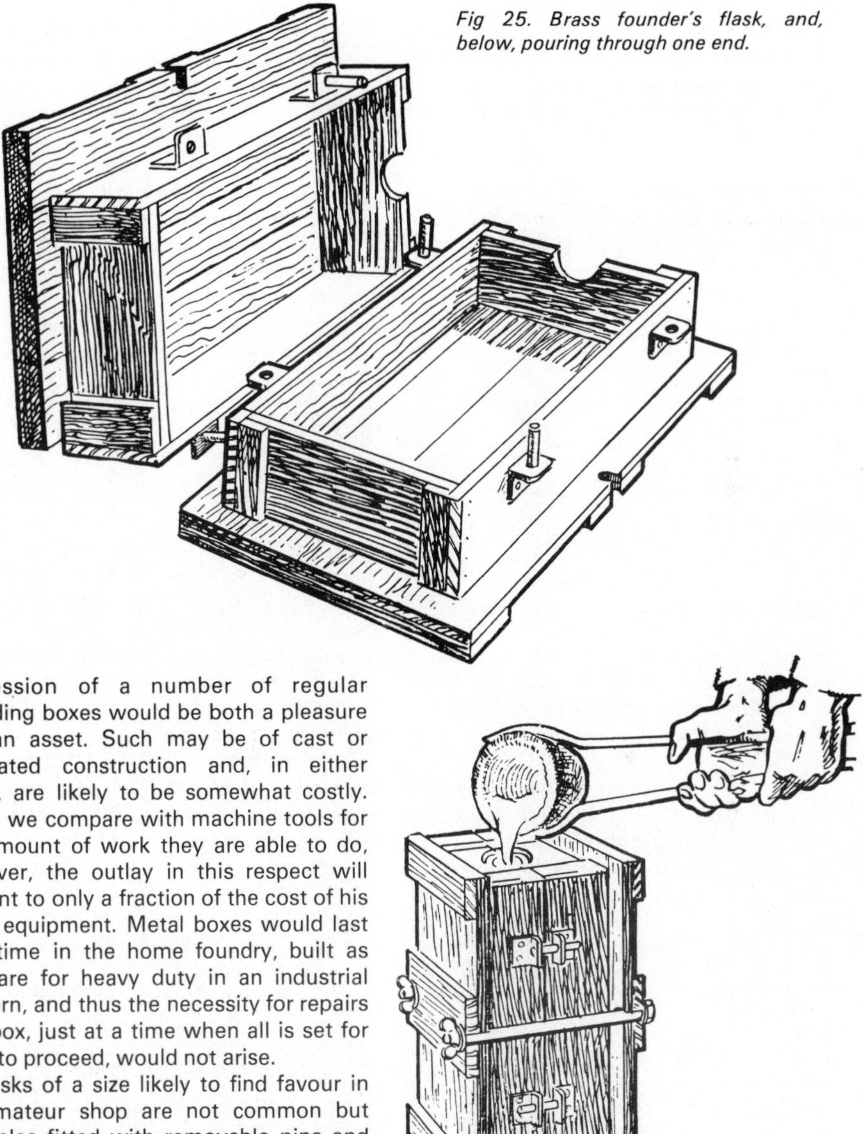

possession of a number of regular moulding boxes would be both a pleasure and an asset. Such may be of cast or fabricated construction and, in either event, are likely to be somewhat costly. When we compare with machine tools for the amount of work they are able to do, however, the outlay in this respect will amount to only a fraction of the cost of his other equipment. Metal boxes would last a lifetime in the home foundry, built as they are for heavy duty in an industrial concern, and thus the necessity for repairs to a box, just at a time when all is set for work to proceed, would not arise.

Flasks of a size likely to find favour in an amateur shop are not common but examples fitted with removable pins and bow handles are supplied in sizes from 8″ × 6″, with a minimum standard depth of two and a half inches.

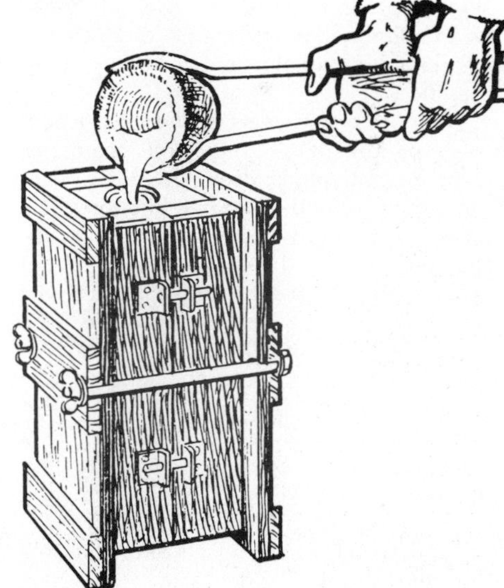

CHAPTER 6

Pattern Making

Materials

Industrially, and doubtless also at the amateur level, a variety of materials can be employed for the making of foundry patterns. Metal, wood and plaster to mention three, and there is little doubt that synthetic materials such as plastics and fibre glass can now play a part. For the purpose of this book, however, the wood pattern is regarded as the basis of the amateur foundry, for even to those with a limited knowledge of carpentry the material is simple and convenient to handle.

In more recent years, following the post war period of permits and scarcities, timber of varying kinds is now freely available. Balsa-wood, of course (though it is a little soft for pattern making), resin-bonded ply in thicknesses from about 1/32 in. ramin, obeche and spruce as well as many others, are now available from model shops and "Do-it-yourself" stores in convenient cut sizes and innumerable sections. At home, i.e. in England, at least, the patternmaker can choose the exact thickness of timber he requires for the job in hand, usually with a fully planed finish.

For a pattern intended for a long "run" a hard wood would be chosen. Patterns for the usual "model size" jobs can, however, very conveniently be made from *obeche* using a "model-aircraft" technique and balsa cement. This timber, intended possibly for model boat builders, is available, like balsa, in strip and sheet, can readily be carved and turned if necessary, cements easily and firmly and can largely be cut with a knife and a rule. Sheets of obeche as purchased have a straight edge and a smooth surface, on which the design of the casting can be drawn with great accuracy. Relative points on a flat surface can be located with confidence, allowing, of course, for shrinkage where the size of the casting demands an allowance. Where small radii are required these can often be formed satisfactorily from the cement itself and small details and even lettering can be added with great facility. Cellulose dope sanding sealer produces a good finish and hardens the surface.

This is indeed "table-top" carpentry. A fine saw and a craft knife are about the only tools needed, although a small plane whose blade has been carefully honed and "stropped" to a razor edge can often be an asset. A jig-saw, a fine bandsaw (as luxuries) or a simple hand fret-saw may be needed for cutting intricate shapes in the thicker sections of wood. Modelling pins

Fig. 26. Pattern making for small engineering projects can often be reduced to table-top proportions – in the manner of model aircraft construction.

and even ordinary laundry spring pegs can be pressed into service for holding parts together while the cement sets.

The maker of miniature foundry patterns will also find useful a selection of ordinary dowels up to, perhaps, half-inch or three-eighths inch diameter. Uses for these are described later in the chapter.

As far as the amateur is concerned the golden rule is, probably, to keep all patterns as simple as possible. The foregoing, however, should not be taken to mean that the results of his efforts should be crude and uninteresting. On the contrary, there is no reason why a model built from home-made castings should not be just as handsome as the full size prototype. Being so very much smaller in dimensions, however, a part may often be produced quite solid, for example, where the corresponding part in full scale may be elaborately honeycombed with cores. If a core is *essential*, use one! Otherwise, make the pattern plain.

If a saw is used, an extra fine blade makes the ultimate finishing of the cross-cut edges a good deal easier and this is usually important. The edges of the pattern are frequently the parts which must later be drawn from the sand.

A knowledge of carpentry is hardly necessary for this kind of pattern making. Indeed, almost any method is permissible providing the finished object is accurate and mouldable. One essential is good finish; not so much from the point of view of appearance as from that a smooth surface will leave the sand cleaner. More attention must be paid to some parts of the pattern, in this respect, than to others

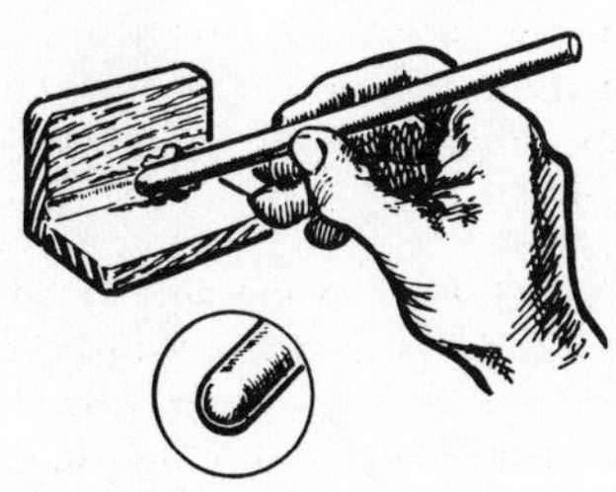

Fig. 27. A round nosed tool employed for shaping plastic wood fillets.

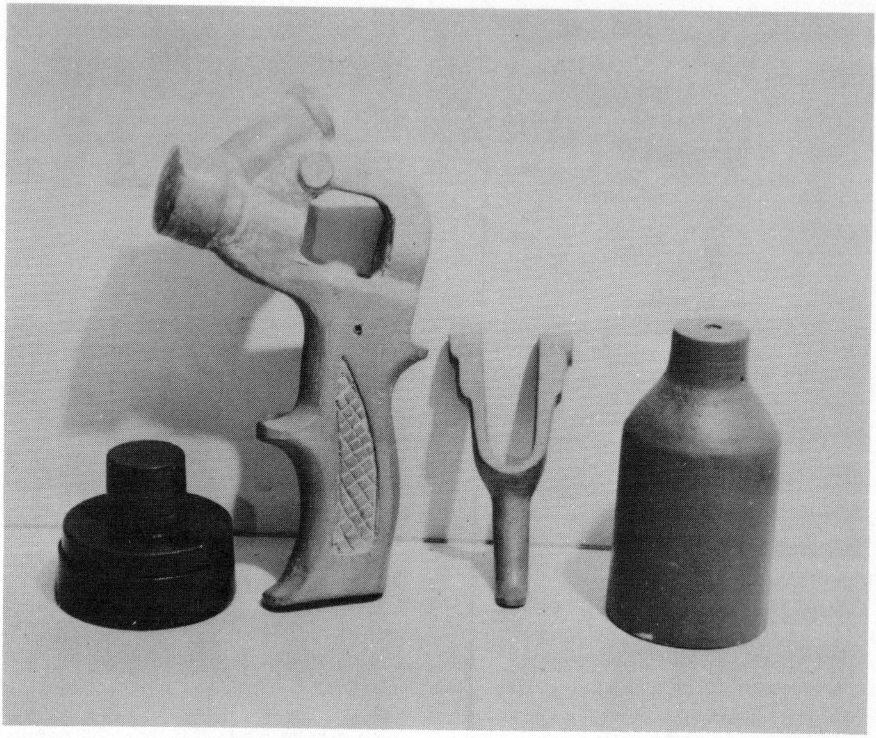

A set of patterns for a spraygun, including the paint container.

and in this matter experience teaches. Often a pattern which fails to draw cleanly at first can be adjusted and smoothed in difficult places until it is quite satisfactory.

Plastic wood can be used to very good effect in forming radii, building up details and levelling hollows and a good, strong glue or cement is indispensable. Ordinary dressmakers' pins with the *points* snipped off are often less likely to split the thin wood than even the finest gimp pins.

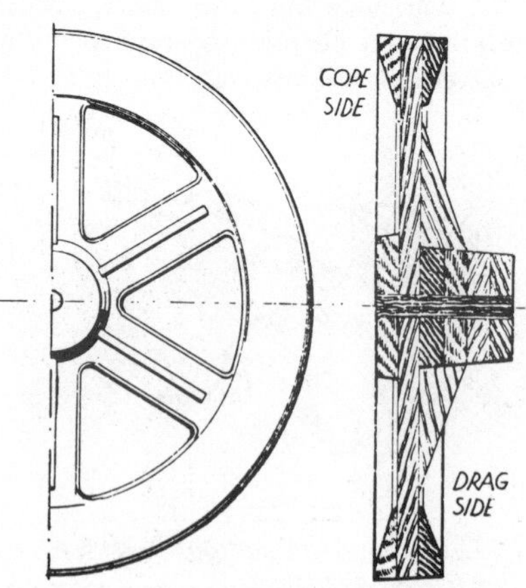

Fig. 28. Construction of pattern for V pulley.

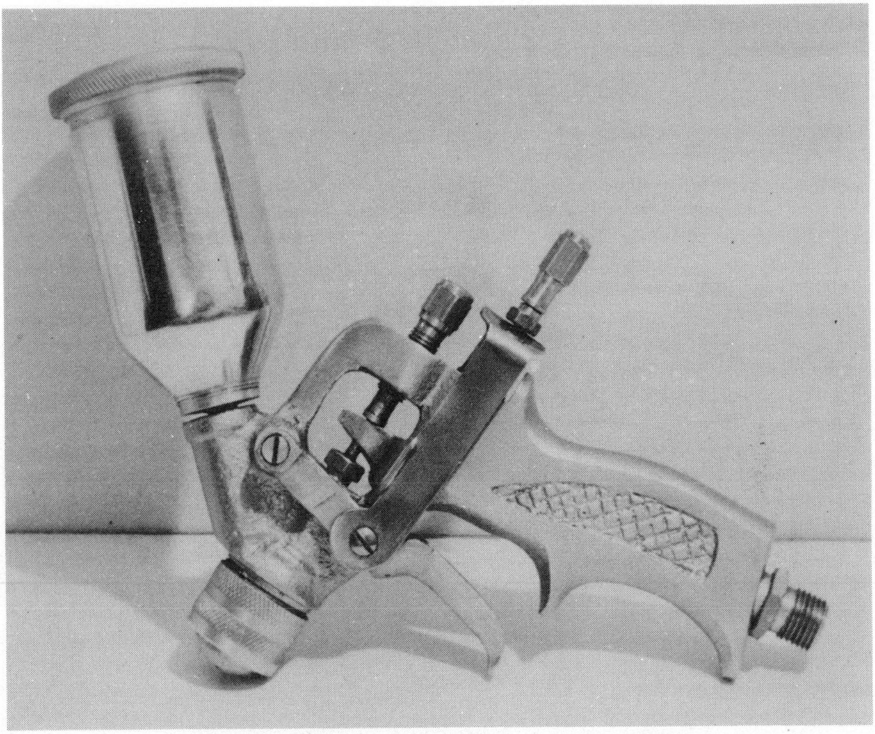

The finished spraygun. All castings are in aluminium.

Construction

Some typical examples of amateur pattern making are given in Figs. 28 to 42. The first is the construction of a pulley wheel pattern, Fig. 28. It is built up from laminations and it will be noted that the bulk of the detail is on one side of the wheel. The feature is not an essential one but it does, materially, affect the ease with which the pattern can be moulded. The point is that the more complex side of the pattern is

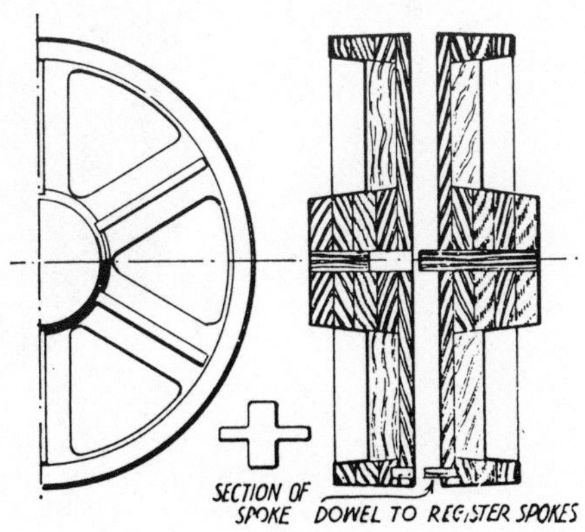

Fig. 29. Split pattern for flat belt pulley.

Patterns for a four-inch cone pulley and a six-inch V-pulley.

Fig. 30. Pattern for cylinder of small locomotive.

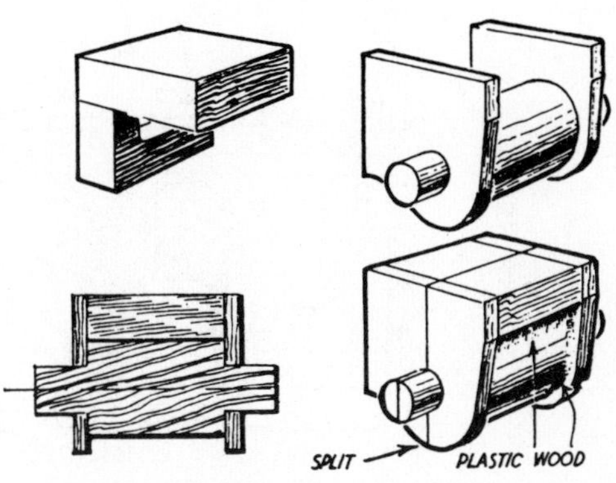

embedded in the lower box leaving practically a clean draw when the upper box is lifted off. It is the kind of thing which can be embodied in the design of a model at the drawing board stage. The alternative would be to split the pattern, i.e. separate the middle laminations as in Fig. 29, a wide pulley with "cross" section spokes. The two parts can be held together in the mould by means of the central peg and the extra dowel to locate the spokes.

It will be seen that, in the case of the former pulley, no provision has been made for casting in the groove, which would require the use of a core. It is anticipated that such a pulley would be employed with an endless vee rope and, that being so, it would be desirable to turn the groove in the casting accurate to the 40° angle and concentric to the bore.

Very often, where a pattern is required for, say, a steam engine cylinder or a similar compact object, the natural tendency is to make it from a solid piece of wood and carve out the detail. Nevertheless, when one becomes used to the idea of the built up pattern, this will often prove the more satisfactory method, as well as providing for construction of a much more accurate character. Fig. 30 gives constructional details of a cylinder suitable for a miniature locomotive. The base of the cylinder, where it is secured to the frames, and the valve face are pinned and glued at right angles. The cylinder itself can conveniently be turned to the diameter required, which will leave just the correct amount of metal round the bore. The bobbin is turned to the same length as the base and each end flange is impaled by a coreprint. When the pieces are assembled as the section shows, plastic wood is applied at the places indicated and the pattern is split through with the aid of a fine saw to separate the halves for moulding.

Fig. 31 shows the construction of a finned cylinder pattern for a 30cc. two stroke petrol engine. The main bulk of the pattern is first built up of laminations sawn out to the cross section. Thus, the flange at the base is a square piece, the cylinder below the finned portion is a series of small discs and the upper part is built up of larger discs. Other details, such as transfer passage and exhaust and inlet stubs are added afterwards, when the pattern has been trued and the fins turned on the lathe. The whole is then split down the middle with a fine saw.

The number of discs required, of

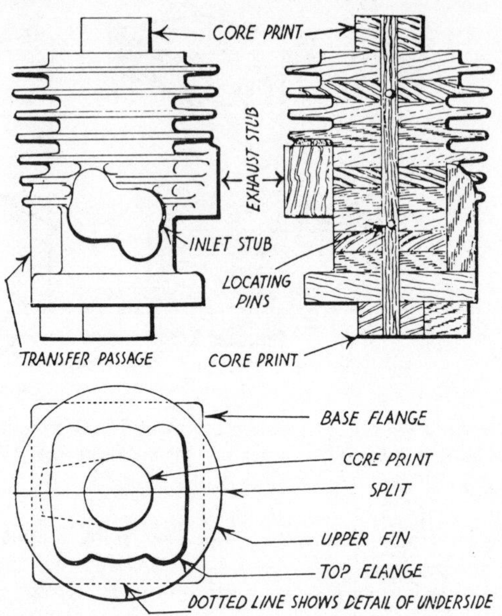

Fig. 31. Laminated construction of pattern for cylinder.

course, depends upon the thickness of the timber and, for the finned portion of the cylinder, this will be most convenient if it is equivalent to the distance between the fins. If the laminations are firmly glued under pressure, however, and allowed to set before turning, the above point is not important. Each lamination is drilled through its centre, $\frac{1}{4}''$, and the whole lot glued and bolted together. A convenient form of bolt, where the turning is to be carried out on a metal-working lathe, is shown in Fig. 32. The taper of the fins is somewhat exaggerated in the drawings.

After the turning has been completed and the fins cleaned up with glasspaper (they require very careful attention in this respect) the bolt is removed and a length of dowel glued in its place. Details can be affixed and plastic wood applied, where necessary, to form radii. A junior hacksaw is useful for slitting the pattern and, if a couple of holes for dowel pins are drilled clean through the cylinder beforehand,

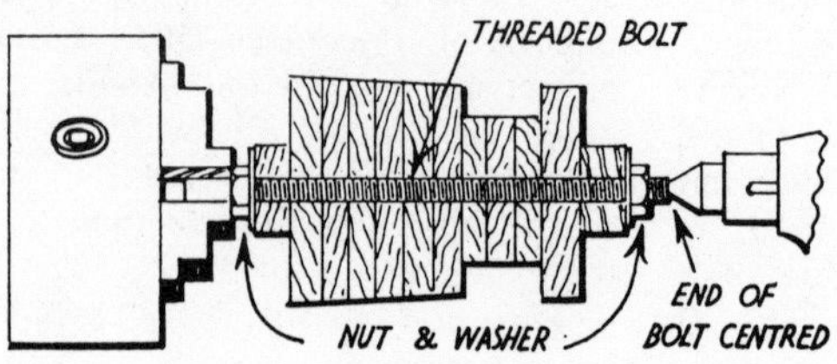

Fig. 32. Pattern threaded on a bolt for turning fins.

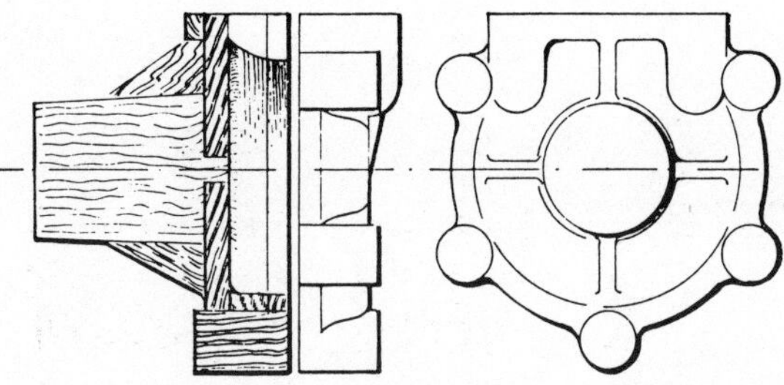

Fig. 33. Pattern for crankcase with overhung crank.

there will be no difficulty about obtaining an accurate register of the two halves. When the pins have been fitted the holes in the outside of the pattern can be stopped with plastic wood.

Another form of pattern which can prove quite a problem is that for the light alloy casting of a crankcase. When the laminated form of construction is applied, however, it resolves into quite a practical proposition. One example is given in Fig. 33 and, here again, it will be seen that the pattern is built up from a series of rings and discs. The two rings, which form the centre flanges between the two halves, are best cut out together. This will ensure a perfect match when the castings are ultimately machined. In the case of smaller models it will be found convenient to cut the middle section of each half from a single thickness of wood and to set over the table of the saw to provide the "draught" angle and taper. Fig. 34.

Very important on a crankcase is the provision of the bolt "pads" on the periphery, through which pass the bolts for retaining the two halves. Quite obviously, they must be in line with each other on the pattern if a good deal of trouble, if not complete disappointment, is to be avoided when the castings are paired up for machining. One suggestion, which will guarantee results, is to clamp the halves of the pattern together temporarily, and add the pads in the form of appropriate lengths of dowel of the diameter required. Fig. 35. These can be secured with glue and pins and, when hard, sawn through to separate the parts of the pattern.

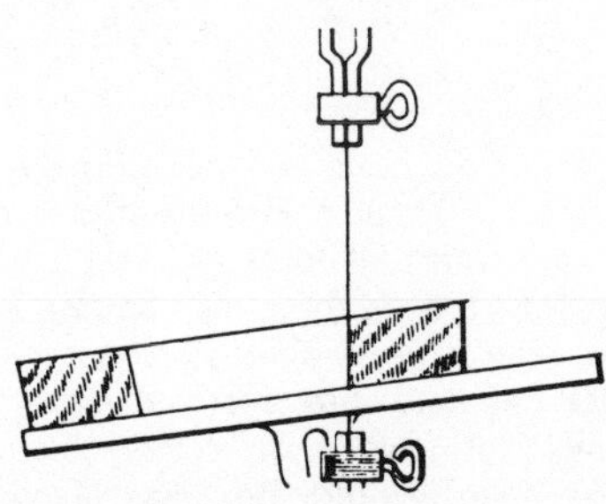

Fig. 34. Saw table set over for forming draught angle.

With crankcases of larger engines, up to 30 or 50cc capacity, it is quite a practical proposition to core out the interior fully, using either self cores, where

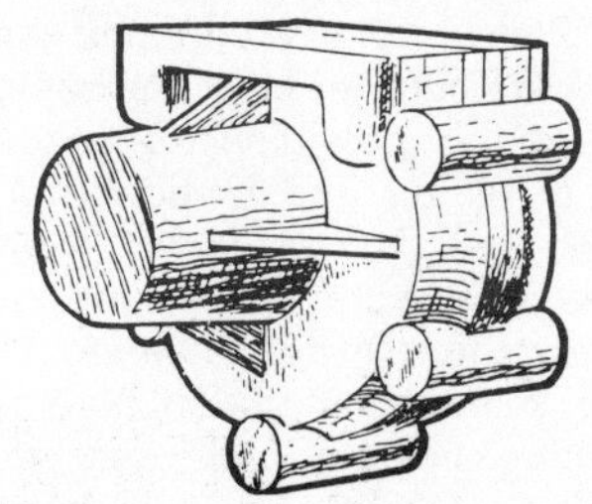

Fig. 35. How to obtain alignment of bolt pads.

Fig. 36. Details of pattern for crankcase.

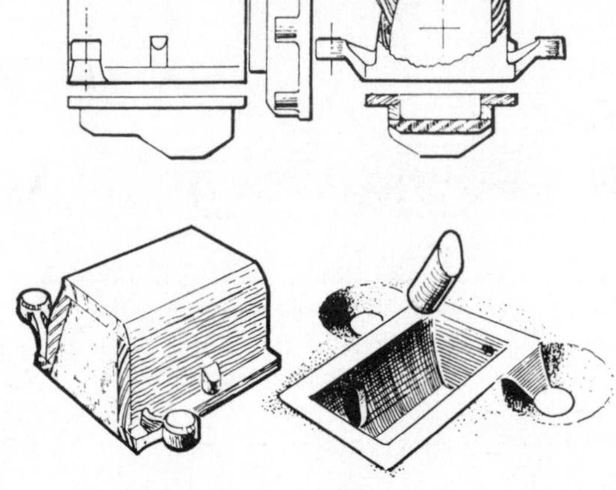

the pattern is hollowed accordingly and leaves a greensand core in the mould itself, or separate cores of baked sand. Wherever practicable, of course, the former plan is preferable. With engines of smaller capacity, however, where a wall thickness of say, three sixteenths or a quarter inch would be excessive, the crankcase pattern can be made either partly cored or quite solid. The wall thickness may then be reduced to as little as a sixteenth, if required, when the casting is subsequently machined.

Fig. 36 shows the pattern of a crankcase for a twin cylinder, four-stroke engine. In this case the pattern is made hollow to leave its own core and the use of a separate core, to clear the camshaft, is avoided by providing for camshaft clearance to be machined from the crankcase wall. The flywheel housing is shown as a separate pattern designed to incorporate the rear main bearing. In this case the main construction is box-like, with engine bearers and other details added. The oil filler stub is shown as a loose part to be withdrawn from inside the pattern before rapping. The indentations in the sand on either side of the pattern show how the parting line has been adjusted to take care of the bearers.

As a change from the laminated and built-up construction of a pattern, Fig. 37 shows the pattern of a handwheel, of quite modern design, which has been carved almost entirely from the solid. As

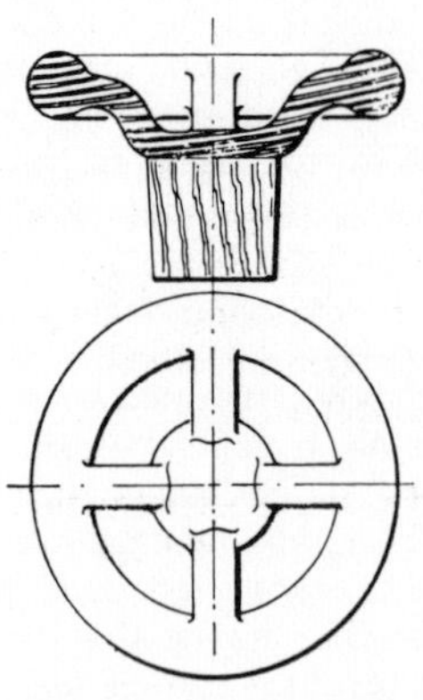

Fig. 37. Handwheel pattern.

A home cast 20cc horizontally opposed glow motor. Castings were in light alloy (melted down pistons) and an iron tube for cylinder liners.

the centre of the wheel is well "dished" this probably represents the simplest method. Here the wheel is, first of all, turned to section on the lathe and the spokes drawn on and pierced afterwards. When turning on a metal working lathe, a beheaded woodscrew is used, held in the three-jaw for chucking. The extended boss is the only addition.

Fig. 38 is the flywheel for a horizontal steam engine and, in this case, the laminated form of construction has been used again. A different feature, however, is the way in which the centre disc, comprising the spokes, has been built up from segments to avoid a cross grain in any of them. The method is quite as effective and a good deal more simple than the

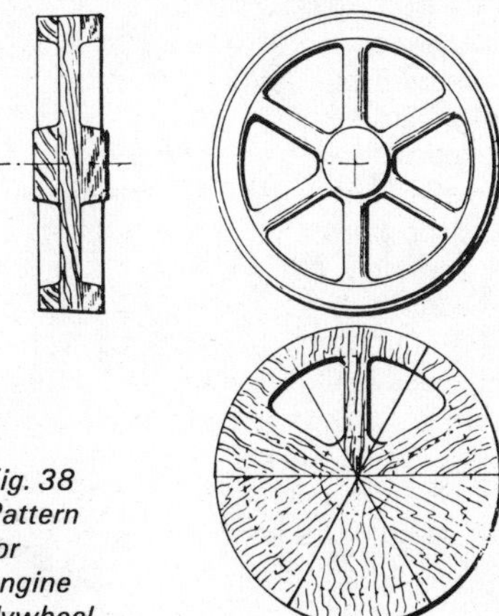

Fig. 38 Pattern for engine flywheel.

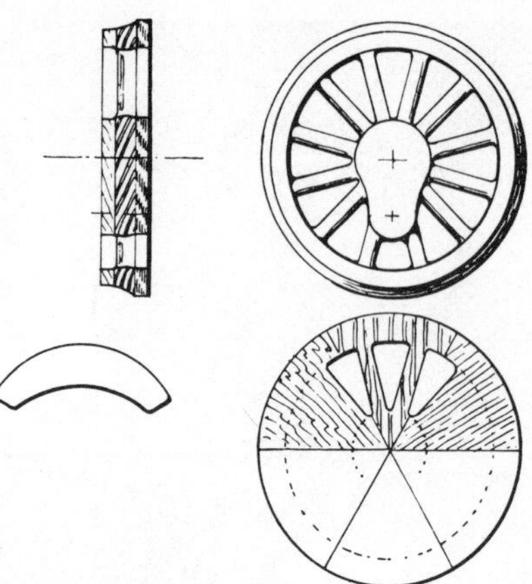

Fig. 39. Pattern for loco wheel.

assembly of a wheel from separate spokes and felloes. The segments are cut out first and glued edge to edge. The rim is formed by the addition of a ring on either side and the hub by two discs in the same way. The outline of the spokes is drawn on the segmented circle after the glue has set and they are cut out as if from a single piece of wood. The spokes are afterwards carved and sanded to oval section.

A similar technique is employed with the next example, a locomotive wheel shown in Fig. 39. In this case the spokes are oval only at the front; the back of the wheel is flat. Balance webs can be varied on different pairs of wheels, if necessary, by the use of loose pieces. Two are cut similar to the segment shown and they are clamped together at the back and front of the wheel by means of wood screws, passed between the spokes. Paraffin wax, or modelling wax, is used to fill in and radius off, temporarily, the spaces between webs and spokes.

A quite different form of construction is illustrated in the drawings of the drill headstock shown in Fig. 40. The basis of this is a piece of wood, three-sixteenths of an inch thick, sawn to the outline at A to form the web. It has projections, a, b and c, which are inserted and glued into slots in the components B and C. Each of the

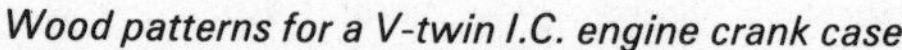

Wood patterns for a V-twin I.C. engine crank case

latter is turned to include core prints in the one piece. Two sets of the ribbing are cut out, simultaneously, on the jig-saw and one is glued on either side of the web. The boss, D, is built up of discs. The example shown was made as a solid pattern. Moulding would be greatly facilitated by making it as a split pattern, although this would entail a good deal more work in the pattern making stage; probably unjustified if only a very limited number of castings were required.

For repetition work, patterns of this nature are usually made as "plate" patterns. That is, a half pattern is mounted to register exactly, on either side of a flat, thin board. The description "plate" is perhaps more applicable to metal patterns made in this way and associated, usually, with mechanical methods of moulding. The difficulty presented by the oddside is, however, avoided in this way and such a pattern is loosened, after ramming up, by rapping the exposed ends of the plate before opening the box.

As it is anticipated that few amateur foundries will normally handle repetition jobs, the solid pattern will probably prove, to all intents and purposes, quite satisfactory. Very often the more difficult

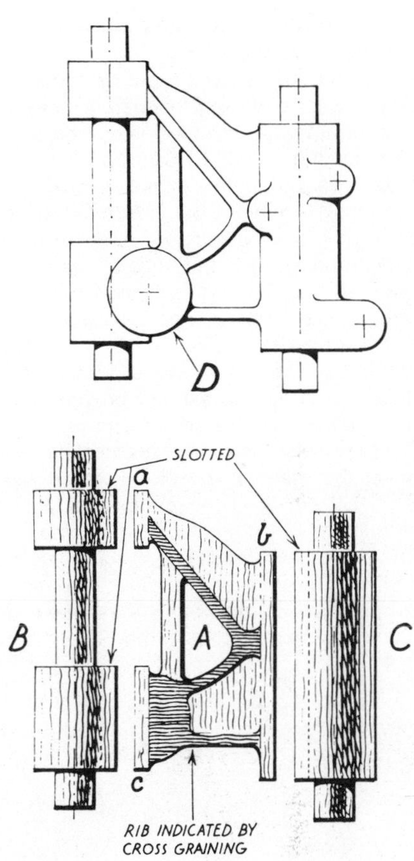

Fig. 40. A drilling machine headstock pattern.

Fig. 41. A template used for moulding pattern.

"oddsides" (the word is used here to describe the *type* of pattern, whereas it would be more correctly applied to the moulding technique), can be moulded more easily by the use of a rough wooden template such as is shown in Fig. 41. Here it is used in conjunction with the pattern for the body of a spray gun, which is itself a particular illustration of the value of the amateur foundry. An illustration of the same gun in use for spraying moulds is depicted in Figure 15.

The pattern is placed on the moulding board and made approximately level by packing up with small strips of wood, etc. The template is fitted round it and this, also, is packed up in a similar way to bring it roughly level with the centre line of the pattern. The inverted drag box is superimposed in the usual way and rammed up. When it is turned over, the template is removed, and the parting line fettled in with a trowel, bringing the mould to the usual stage where it is dusted with parting sand or powder, prior to the addition of the cope.

Large cylindrical patterns can be built up in the manner shown in the drawing (Fig. 42) which is the pattern for a cast aluminium petrol tank, such as may be fitted to a lawn mower or a stationary i.c. engine. The domed ends are a series of laminations as previously described for many of the other patterns. The centre section is built up on two hexagonal formers and, afterwards, skimmed off on the lathe to the diameter required. Filler neck, outlet boss and bracket for mounting are added afterwards and all interstices are filled in with plastic wood. The pattern is finally split with a saw; the two halves, afterwards, being retained with dowel pins.

Contracting on Cooling

A point about patterns of which no mention has yet been made is the normal procedure for providing an allowance in the measurements to take care of contraction in the casting when the metal cools. In the ordinary way a casting in aluminium will contract at the rate of up to a quarter inch to the foot of pattern according to the section, iron at the rate of one eighth inch to the foot, with most other alloys somewhere between the two. It will be appreciated that where castings for model making and so on are only a fraction of a

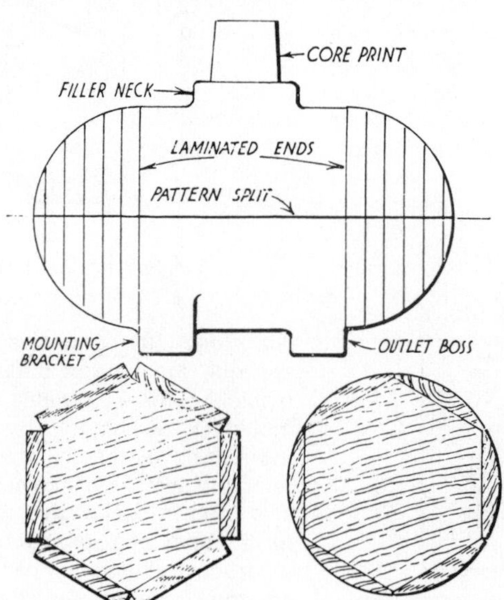

Fig. 42. Example of a large cylindrical pattern.

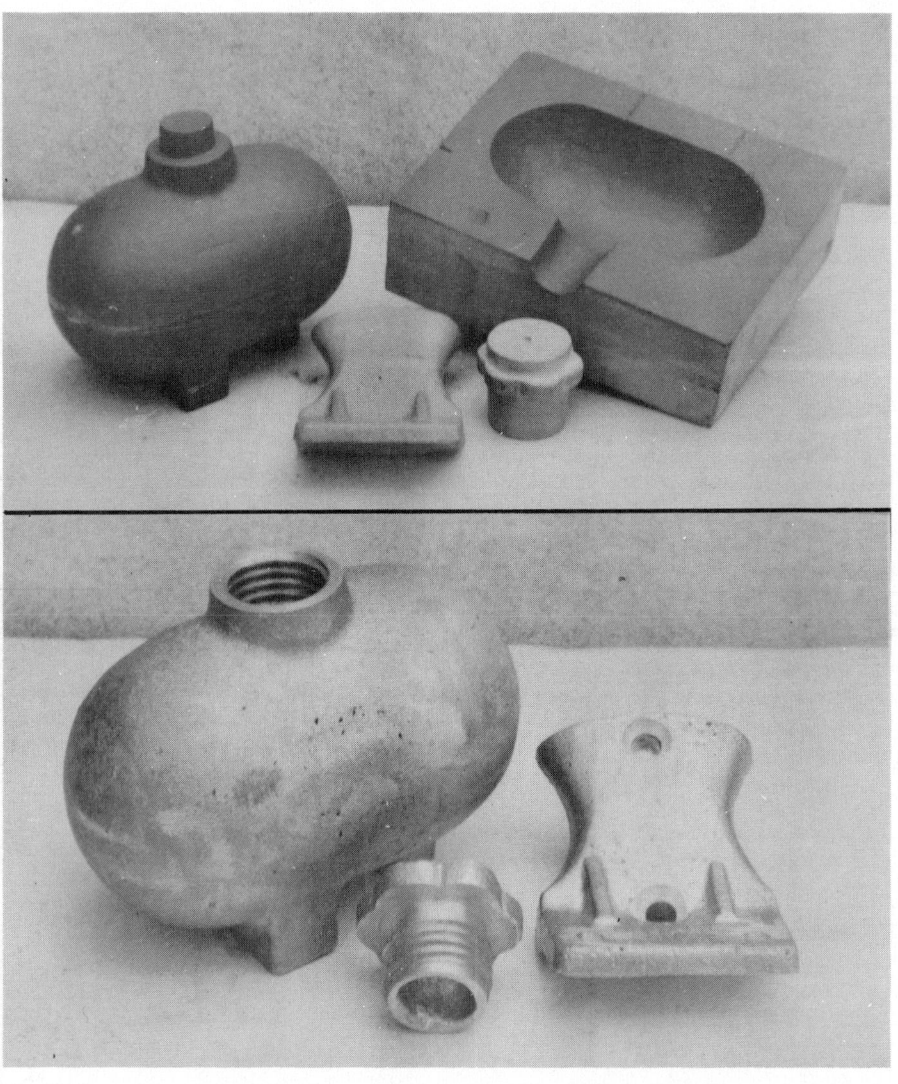

Top, quite an awkward pattern to cast. A description of the technique appears opposite. Lower photo shows castings obtained from the patterns to produce an aluminium fuel tank.

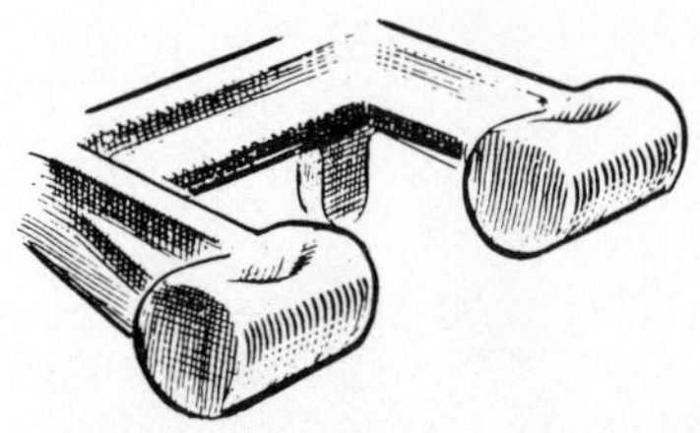

Fig. 43. Casting spoiled by local contraction.

foot in any measurement, the actual allowance due will only be quite small. Frequently, therefore, it will be found that, where a casting is to be later machined, the allowance for machining, if made on the generous side, will be sufficient to take care of shrinkage as well. Obviously this dismissal of the problem will be governed to a great extent by the form and design of the casting contemplated and more attention will have to be given to the matter when a casting has, say, two or three bolt bosses, to be accurately located, spaced some six inches apart. A linear contraction allowance would then be a necessity, but as a rule, the worst effect of contraction that the home worker will encounter is the kind of *local* shrinking or dimpling which sometimes takes place in the thicker sections of a casting.

Fig. 44. Building up a section with plastic wood.

The fault becomes manifest in castings in aluminium and cuprous alloys where it is a problem which confronts professional and amateur alike. The shrinkage must take place somewhere when the metal cools and it should be the policy of the moulder, where possible, to cause it to occur away from the main body of the casting. The provision of a down runner of large diameter is a contribution in this direction. While the outside of the metal in the runner solidifies the centre remains liquid long enough to "feed" the casting. Methods of feeding, however, belong to the chapter on moulding. As far as the problem concerns patternmaking it may be of some assistance to build up such doubtful parts of a pattern to make *allowance* for the shrinking. A slight ovality on the upper side (as cast) of a cylindrical section of pattern, for example, will often prove an answer. Fig. 44. In the case of wheel hubs and other heavy bosses cast with a vertical axis and without coring the remedy is, often, simply to extend the boss by the thickness of, perhaps, one lamination. Thus the increase, with contraction flaw, can be machined away afterwards.

CHAPTER 7

Cores and Coremaking

Reference has been made a number of times to cores, coreboxes and coreprints. The following chapter deals with the kind of core which is made quite separately from the mould and is placed in the cavity after the moulding process has been completed. In sand moulding and even, occasionally, in some classes of die casting, where a permanent, metal mould is used, cores are made from sand of a special mix, such as has already been described under the heading, "Oil Sand."

Sand cores are produced in different ways. By extrusion, for example, by moulding in a corebox and by building up over a metal former. In the class of work undertaken by the amateur foundry, however, the moulded type of core will almost certainly predominate. A corebox in, perhaps, its most elementary form, is shown in the illustration, Fig. 45. This is made in wood, in two parts, which are registered together by the pegs A and A, and produces a straight, cylindrical core such as may be employed in moulding a steam or petrol engine cylinder or for the drill headstock shown in Fig. 40.

Two rectangular pieces of wood are cut, of a length equal to the length of the core required, and of a depth to allow for gouging out the semi-circular troughs. The two rectangles are matched and drilled through at the points indicated for fitting the pegs. A circle is scribed, of appropriate diameter, at either end while the halves are registered; separated, the circles are connected by ruled pencil lines across the face of the wood. The template B is used as a guide to gouging.

In use, the two halves of the box are brought together and held by the fingers of one hand (or, in the case of large or ungainly boxes, by means of a clamp), while the coresand is pressed in with the other. The box is held, vertically, on a flat surface, for ramming, so that the lower end is closed and the sand is fed in little by little and packed down progressively. To release the core, the box is rapped vigorously on either side when it can be lifted or rolled out on to the drying plate.

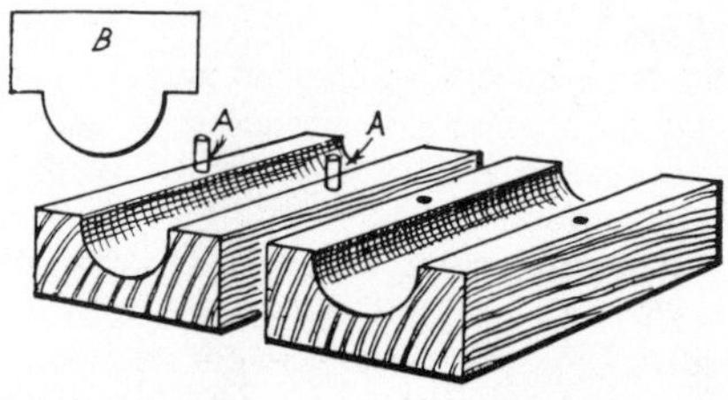

Fig. 45. Simple corebox with template.

When they are baked, small and simple cores will be strong enough to support themselves and, generally, they will be sufficiently permeable to allow for the escape of the gases. Nevertheless, the habitual reinforcement and venting of *all* cores, large or small, intricate or simple, can be a valuable insurance against possible failure on either count. Reinforcement takes the form of wires, or even nails, either placed in the box during

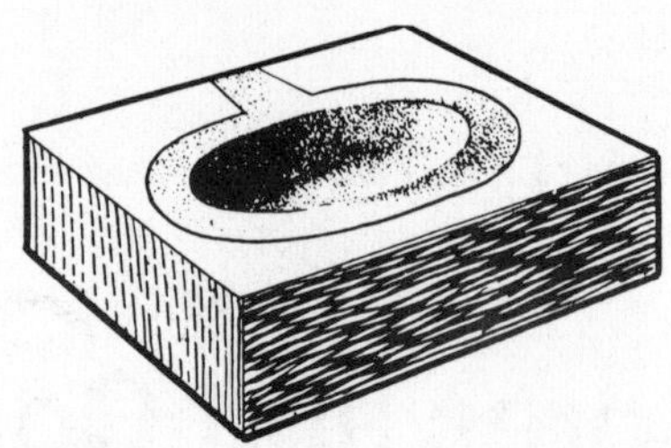

Fig. 47. Large sand core made in half corebox.

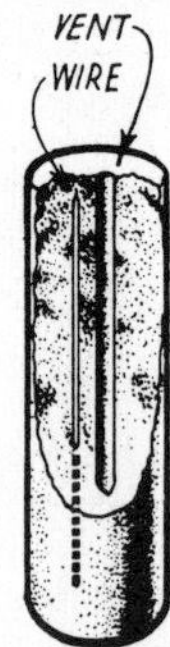

Fig. 46. Showing how a sand core can be vented and also reinforced with wire.

ramming or pressed into the core afterwards. The core can be vented with the aid of a pointed wire, about an eighth of an inch thick, which can be forced in and then withdrawn. An alternative is a length of well waxed cord embedded in the sand during ramming. On baking, of course, the wax volatilises and leaves a tiny cavity round the cord which, in any case, is probably also charred away. It is a technique which is particularly useful when dealing with intricate or awkwardly shaped cores.

Sometimes cores, which are formed of a large body of sand, and which may be moulded in two or more separate pieces and joined together, after baking, with core gum, dextrine or other adhesive, can be vented by digging a hollow in the deeper sections of sand and connecting this to outlets formed by means as already described. An example of this is shown in the drawing of the half corebox, used for making the core for the petrol tank, Fig. 47. Only a single half box was required for this core, both sides being identical. The half cores thus formed, by virtue of them being made hollow, also afford an economy in coresand.

When a mould is intended for fitting with loose cores, provision for locating them is usually made on the pattern. The core "prints," as they are called, take the form of projecting pieces of wood fixed to the solid pattern at the points where the cored hole is to break through the casting. It is often possible to embody features in the design of a model or machine part which will allow of the cores being adequately supported in this way without having to provide additional prints and corresponding projections on the core, producing holes in the casting other than those actually required. The practice is common in the casting of full size water-jacketed cylinders and the plugs used for sealing the holes are in evidence on the sides of the block. There are occasionally, however, other means of anchoring a core which is otherwise rather insufficiently supported by its prints. For example, in the case of the petrol tank, again, it was

Fig. 48. Use of chaplets for retaining a suspended core.

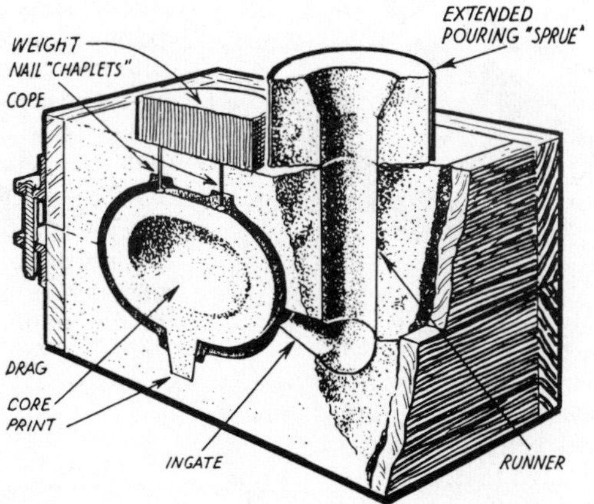

necessary to make use of a couple of nails to prevent the core from floating, the core in this case being virtually balanced upon its central print. The arrangement is indicated in Fig. 48. The use of "chaplets" in this way is quite orthodox practice and various forms are in use in foundries. In the positions in which the nails are placed there is no detriment to the casting whatever. One of them is in the centre of the outlet bush and that will be drilled out for tapping in any case. The other is placed in the centre of the mounting bracket where, in the unlikely event of any leakage occurring, there is ample thickness of metal to allow for the insertion of a tapped plug. The nails are inserted in the mould from the inside and, when the box is closed after fitting the core, the heads come into contact with the latter and evidence of them lifting a little will, in a case like this, be apparent by a slight disturbance of the upper crust of the mould.

Occasionally, the corebox constitutes the most important part of the pattern. An example of this is the piston for an internal combustion engine. The pattern, shown in Fig. 49 is a plain, cylindrical one with a tapered coreprint at the base and a spigot, not necessarily of reduced diameter, at the top for subsequently chucking the casting for machining. It may be made in two parts, retained by a central dowel, as an aid to simpler moulding, but that again is a matter of choice. The core, on the other hand, carries all the internal features

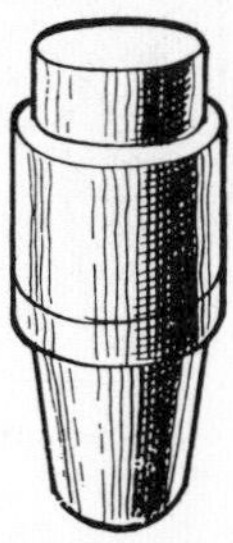

Fig. 49. Piston pattern.

of the piston design according to the size and type of the engine. A piston corebox and core produced are shown in Figs. 50 and 51.

Basically, the box is of the kind illustrated in Fig. 45. Due to the varying diameters of the core and to the tapered

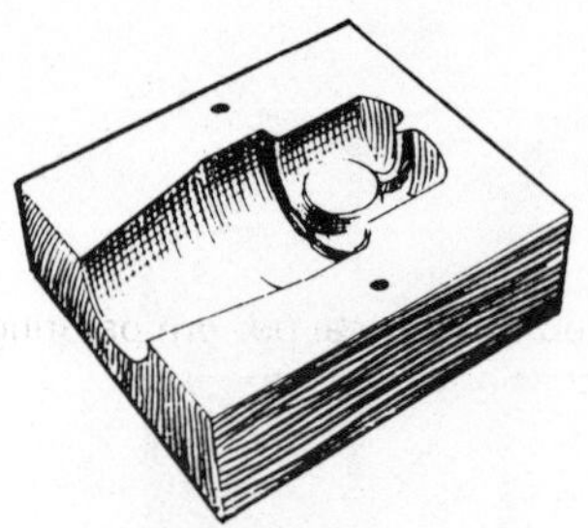

Fig. 50. Corebox for piston.

print, however, more than one template must be used. When gouging out the box it is an advantage to carry the groove right through and to fit the crown details, in the form of a split plug, afterwards. For fitting the gudgeon pin bosses it is desirable to mark their centres accurately and to drill right through the box. Small wooden bosses are then turned, to include a spigot to fit the drilling, and glued in place. Their alignment can thus be regulated with reasonable certainty.

Fig. 51. Piston core.

Another example of a corebox and core is that for a twostroke cylinder in Figs. 52 and 53. In this case, it will be seen, the core embodies the exhaust and transfer passages and, for that reason, must be designed with particular care, as these time the valve events of the engine.

Although the foregoing examples touch only the fringe of the subject of cored moulds, they may serve to indicate the general lines upon which work and experiment can proceed. As a rule, perhaps, the amateur foundryworker will enjoy the greatest success by the moderate use of loose cores, particularly when dealing with comparatively tiny model components, but there will always be the occasion when a core is indispensable to the job or where the saving, at the later machining stage, more than compensates for the additional work of patternmaking and moulding.

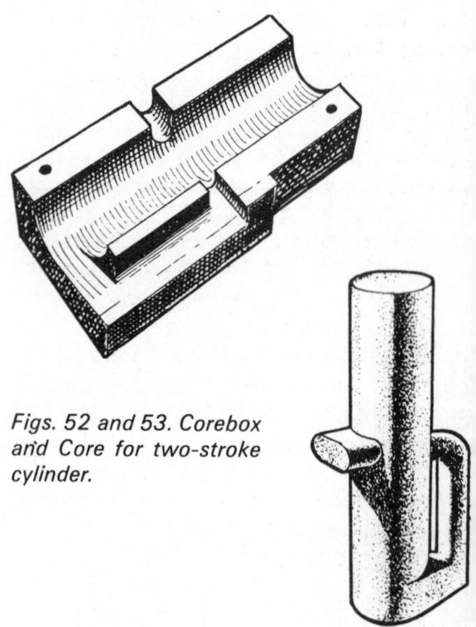

Figs. 52 and 53. Corebox and Core for two-stroke cylinder.

Painting

Before leaving the subject of patterns and coreboxes altogether, it will be as well to deal with the question of painting them. Paint or varnish is applied to a wooden pattern for two reasons. One is to preserve it from the effects of moisture which is deleterious to the wood, raising the grain, causing warping and softening the glue. The other reason is that the paint imparts a smooth, hard and even surface to the pattern which is thus, at one and the same time, easier to mould and leaves a cleaner impression.

The kind of paint to be used is not important so long as it dries hard and is non-tacky, and, as a rule, a glossy finish is preferred. Oil bound paints, spirit shellac or cellulose are equally effective and the use of one or another by the amateur will usually be dictated by its availability. It is an advantage, and worthy of consideration when choosing, if the paint is quick drying. A slow drying paint can easily spin out a simple job over a couple or three days by the time the wood has been treated; first, to a primer or undercoating, which must be quite dry before glasspapering to remove the "whiskery" effect when soft wood has been used, and then to a finishing coat of gloss, which must again be quite dry and hard before any attempt is made at moulding.

Where it is desired to cut painting time to a minimum there is a good deal to be said for the use of brushing or spraying cellulose – particularly the latter – when a couple of coats can often be applied within the hour. In many respects the material is ideal. Cellulose filler is extremely effective in making the surface of the pattern even. True, if the filler is applied rather too liberally there is a tendency towards surface cracking to appear after some weeks or months but, used as it should be, there is no reason why patterns finished in cellulose paints should not remain as serviceable as any others.

Although the spray can hardly be bettered for general pattern painting there are occasions when the use of a brush is indispensable. The example given in Fig. 54 shows what happens when an attempt is made to spray paint between the fins of a cylinder. Ordinarily, little or no draught is given to them in the first place and, obviously, their surface must be very smooth and unimpaired or the sand will cling. The paint here has built up round the outer edges of the fins giving them what amounts to a *negative* taper and making a clean strip impossible. For work of this nature there is an advantage in the use of a soft sable or squirrel brush. Brushes very suitable for all small painting jobs are those which fall in the category of "writers" and which are usually obtainable in a variety of flat and round sizes from artists' colourmen, or from the kind of good oilshop which caters for the requirements of a professional decorator.

Fig. 54. "Draught" on finned cylinder pattern spoiled by globules of paint.

CHAPTER 8

Making a Greensand Mould

Arriving now at the stage where, let us suppose, a modest but useful collection of foundry impedimenta has been acquired, the amateur engineer finds himself in need of one or two extra items of equipment in his workshop. A small machine vice, shall we say, which would normally run to about £10 to purchase but which, he believes, he can make for himself at a complete saving. At the same time he will be able to embody such ideas or special features as may be to his own liking while nothing need be said for the enjoyment to be derived from the pursuit. It is also true that he can often make his own castings, allowing even for pattern making as well, in less time than it would take to order by post from a supplier.

The vice illustrated in the drawing, Fig. 55, is included as an example and not with any claim to special merit. It would, however, according to the size it was produced, form a very useful addition to the accessories used with any lathe up to the popular three or four inch sizes. Its construction calls for the making of two patterns, one for the fixed jaw and bed of the vice and the other for the moving jaw. The castings are to be made in iron.

Fig. 56 shows the patterns separately and arranged on the turning over board ready for moulding. Both have a flat base and fall into the category of being the simplest type of pattern to mould. They are placed base downwards and spaced so that there is room for them within the dimensions of the flask – indicated here by the dotted lines – with something in excess of half an inch to spare between the nearest points of the patterns and the sides of the box, due regard being paid also to positioning the runner (dotted circle) where its attachment to either casting will not be an inconvenience but will, nevertheless, assure the mould being completely "run." The last point is one which will largely be dictated by the exercise of common sense *not* unallied with experience. Errors in this respect will occur with even the most practised moulder when a pattern is somewhat out of the ordinary but, with the example under discussion, there should be little cause for failure.

The drag box, inverted, is positioned round the patterns. In this case the drag is the part without the pins and thus the box will lie flat upon the board. In some instances the box *with* the pins is filled first, depending on the preferences of the moulder, and this then becomes the drag, but there is no hard and fast rule.

Fig. 55. A useful subject for the home foundry.

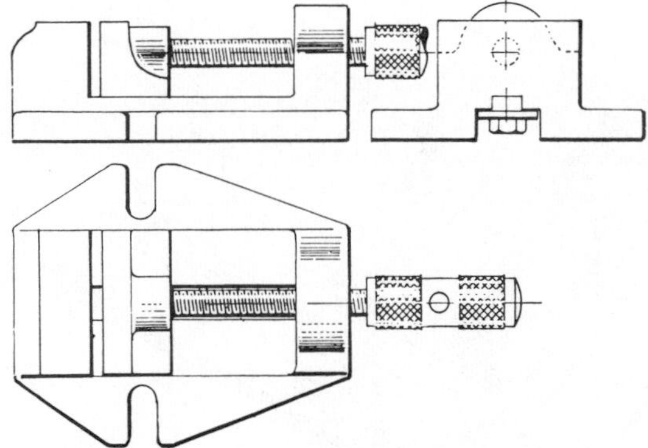

Obviously, in the latter case, allowance has to be made for the pins, either by providing holes to receive them or for them to overhang the board. Some commercial boxes are provided with loose pins and thus there is no problem.

The sand has been prepared by tempering with water and has been sieved ready for use. Ramming commences by covering the pattern and, perhaps, the most effective way of doing this, especially with objects which may present a more intricate contour than the one in hand, is by taking the sand up in handfuls and throwing it, virtually, at the pattern. Deeper hollows and crevices will receive more particular attention to ensure that the sand penetrates fully and with even density. Sand is drawn up to the sides of the pattern by hand and pressed there firmly. Attention to this detail will ensure a smoother finish to the sides of the casting

A cast iron machine vice of the type shown above – always a welcome addition to the tool kit.

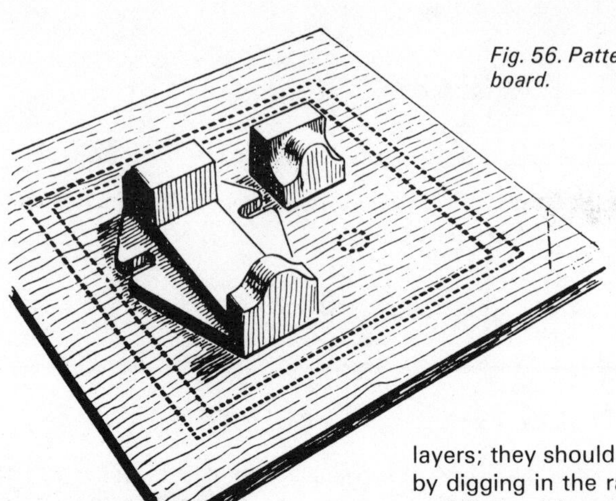

Fig. 56. Patterns arranged on moulding board.

than if the sand is just piled in and allowed to take its own course. Details can be attended to by pinches of sand directed at them until the moulder is satisfied that the whole area of the pattern has been dealt with effectively. It may be well to mention that, in professional moulding practice, "facing sand," i.e., new sand straight from the mill, is usually employed in direct contact with the pattern, the remainder of the box being filled up with used sand, which has been retempered, of course, from the floor of the shop. It will take a long time for the sand of the amateur foundry, if it has been acquired new, to deteriorate to the extent that it is unfit for use against the pattern so, to a large degree, that particular detail can be ignored.

Sand can now be scooped into the box a little above pattern depth and the first layer, around the pattern, treated to a ramming. Quite a useful rammer like that shown in Fig. 58 can be made from a short length of broomstick. Sand is best added in layers and the ramming carried out in stages until the box is filled. Just the same, it is unwise to separate the layers; they should be well keyed together by digging in the rammer or sand is liable to fall out when the box is turned over. The drag is finally tidied up by strickling away the surplus sand and levelling the top, Fig. 59. The packed box is now vented by means of a pointed wire driven in a number of times above – but not so deeply as to touch – the pattern, Fig. 60. Wire of about one eighth of an inch in diameter is suitable and often, a fine "radio" type of screwdriver, which may have outlived its normal function, will carry out the work admirably.

Fig. 57. Moulding by sand slinging.

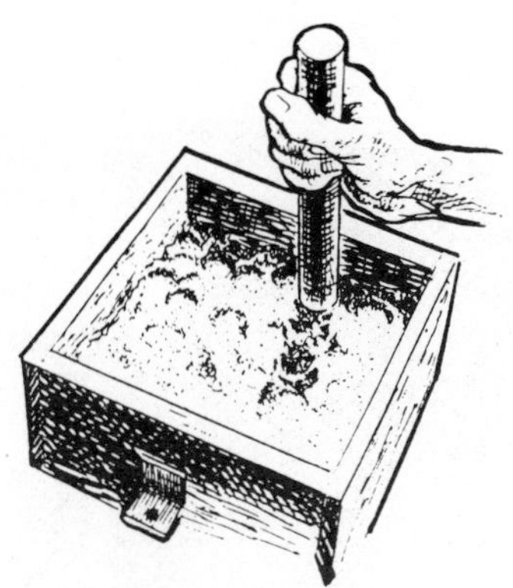

Fig. 58. Ramming.

The drag box is now turned over. If it has been rammed effectively there should be no need to support it in any way to do this. It should be possible to lift a mould quite safely without risk of the sand falling out although, where the area of a box is very great in relation to its depth, it is usual to provide it with reinforcing bars across the middle. The surface now uppermost, revealing the undersides of the patterns, and being the joint face of the mould, is given a dusting with parting sand or powder. Fig. 61 shows the process using a coffee strainer. Plastic ones are quite cheap and distribute the powder very evenly.

Fig. 59. Strickling away the surplus sand.

Before adding the cope or top part the stick or tube for forming the down runner is positioned. For iron, a short length of brass tube, about half an inch in diameter, will produce an excellent runner and the position of this, relative to the patterns, is shown in Fig. 62. The tube is pressed into the sand and will support itself vertically

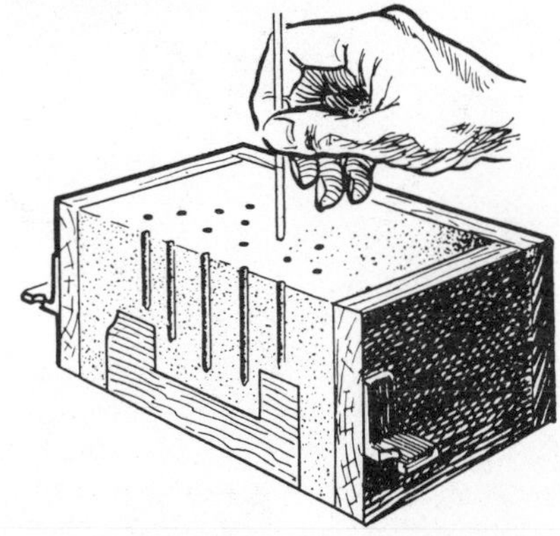

Fig. 60. Venting with the aid of a pointed wire.

Fig. 61. Dusting with parting powder.

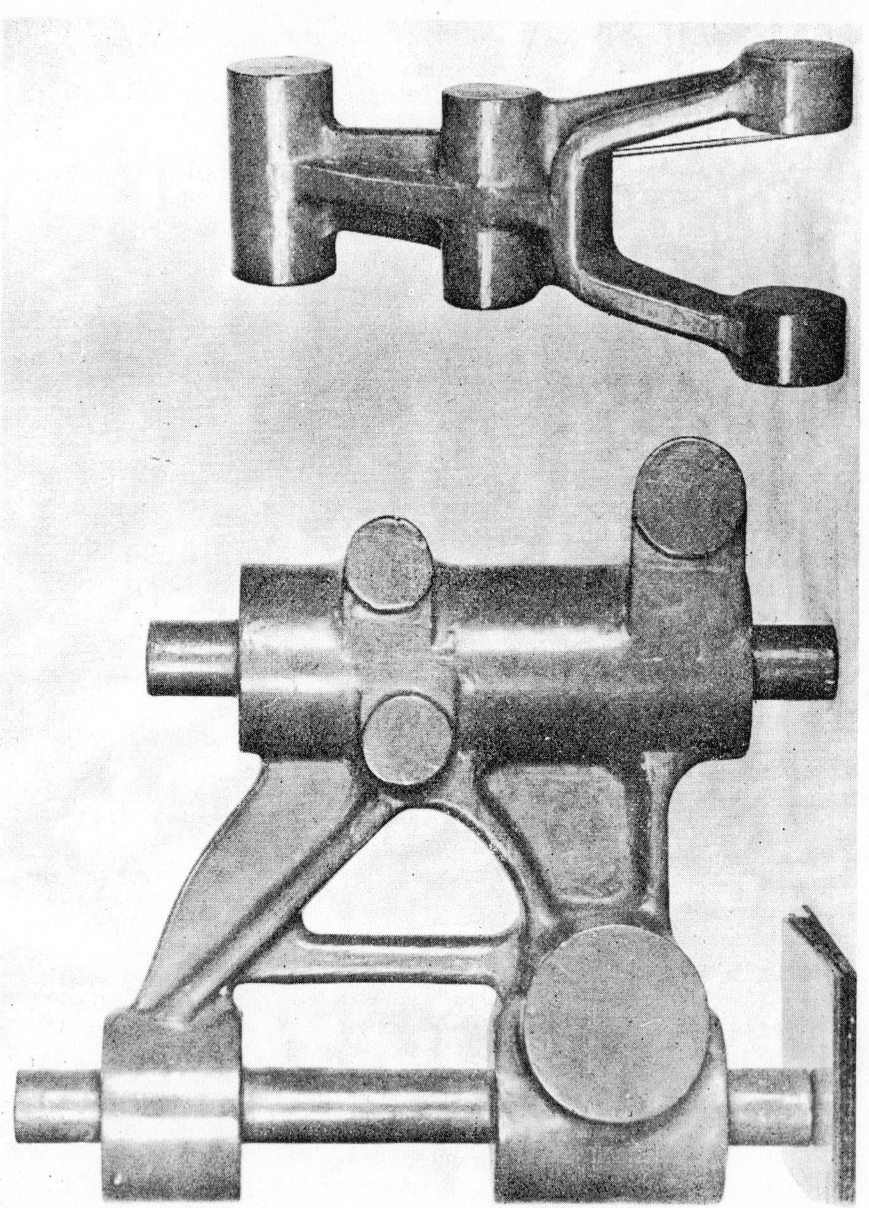

Patterns for a drill headstock and jockey bracket

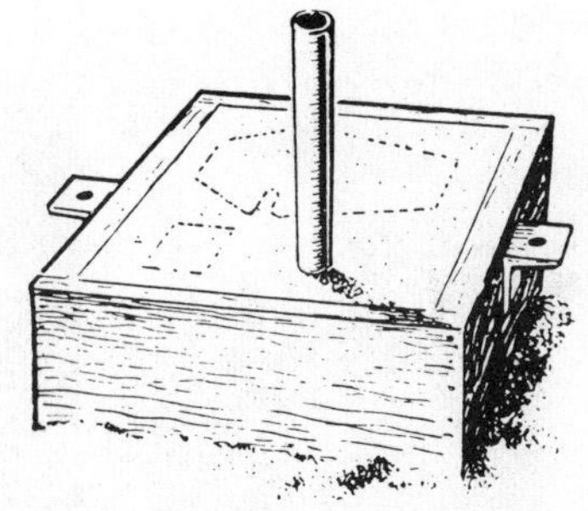

Fig. 62. Positioning the runner tube.

until the cope has been rammed. Some importance can be attached to selecting just the right spot for the runner and much will depend upon the shape of the pattern. It is usual to pour into one of the least important sections of the casting and, in this case of course, it will be into the base flange. The distance of the runner from the pattern should be a compromise: not so close as to cause damage or weaken the mould when the ingate is cut and not so far away as to appreciably reduce the temperature of the metal before it enters. The cope is now placed in position and rammed up in the manner described for the drag. If a finger is placed upon the runner tube in the earlier stages, to steady it, there will be little danger of it shifting while the box is filled. The top is strickled off and vented as before.

Fig. 63. Forming the runner.

The next stage is to withdraw the runner tube. It is loosened by a circular motion, as shown in Fig. 63, which also serves to widen it somewhat. The mouth of the runner is then cut to a funnel shape with the aid of a trowel or moulded with the fingers. The simple operation is carried out with even greater facility by the use, for example, of a small plastic funnel A (Fig. 64) or the bell of a child's discarded trumpet, B. Failing either of these objects being conveniently to hand it may be well worth while making such a tool, by casting a similar shape in aluminium and turning or polishing to finish.

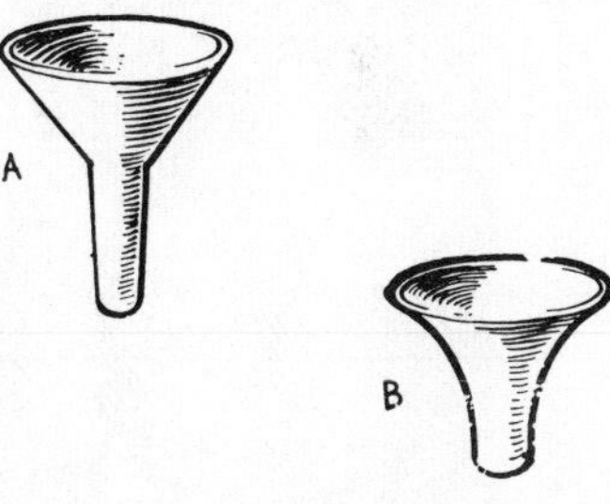

Fig. 64. Two simple devices for widening the runner mouth.

Fig. 65 illustrates the form and comparison of typical runners from, on the left an iron casting and, on the right a casting in aluminium. In the latter case the greater bulk makes for sound castings and a minimum of shrinkage. There is some

The headstock pattern shown on page 60 part buried in the sand of the drag.

compensation for having to pour "to waste" such a large volume of metal in that, very often, the runner can be put to good use in the workshop in the turning of small parts. That is, of course, when it is not required to put back into the pot for making up the next melt. The same can be said of runners from bronze and gunmetal castings which can, quite advantageously, be made of similar large section. Often, if the runner is sound, useful bushes and so on can be turned from these.

The mould is now split and the cope, which in this case requires no further attention beyond dusting or spraying with plumbago, is stood aside out of harm's way. Before withdrawing the patterns from the sand it is usual to moisten the sand immediately surrounding them, using a camelhair brush for applying the water. This is to guard against the possibility of the edges of the mould crumbling as the pattern is removed. Apart from loosened sand falling into the impression to spoil the casting the preservation of the firm edge to the cavity avoids excessive "flash" – the term used to describe the thin, ragged web of metal sometimes seen round an unfettled casting at the junction of the two parts, evidence of an ill-fitting mould.

Now, also, the ingates can be cut to connect the impression left by the runner tube with the edge of the pattern. There is an advantage in cutting the ingates at this

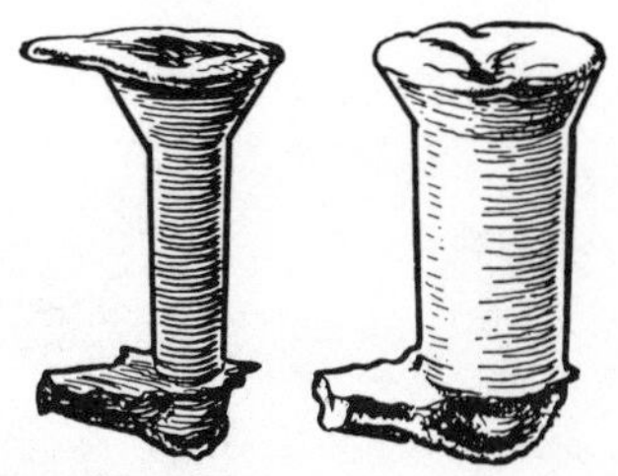

Fig. 65. Comparison of runners from iron and aluminium castings.

stage, instead of after the patterns have been removed, as the risk of damaging the mould is thus reduced. A suitable and easily made tool for cutting ingates is shown in Fig. 66 A. B is a typical combination moulding tool. One end cuts ingates, etc., while the other is a small trowel of much general utility. In a home foundry, where a casting session occurs at irregular intervals, often with an appreciable lapse of time in between, moulding tools are best made of brass. Steel tends to rust all too quickly unless in continuous use. The form to which the ingate is cut is illustrated in Fig. 67. The tapered section of it allows a fair body of metal to be brought close to the cavity without losing heat and the rather narrower "throat" makes fettling easier. If the runner is to be broken off it will snap precisely at the thinnest part or, if to be cut through with a hacksaw, there is a conveniently small section for the purpose. The dimensions of

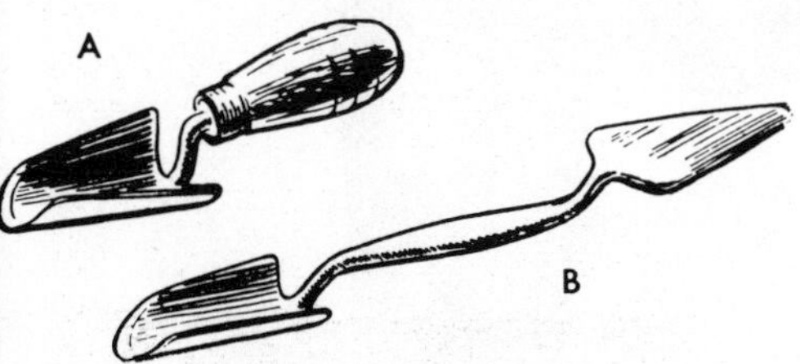

Fig. 66. Tools for cutting ingates.

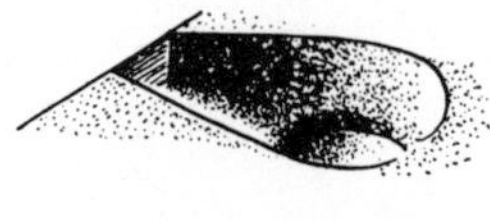

Fig. 67. Ingate.

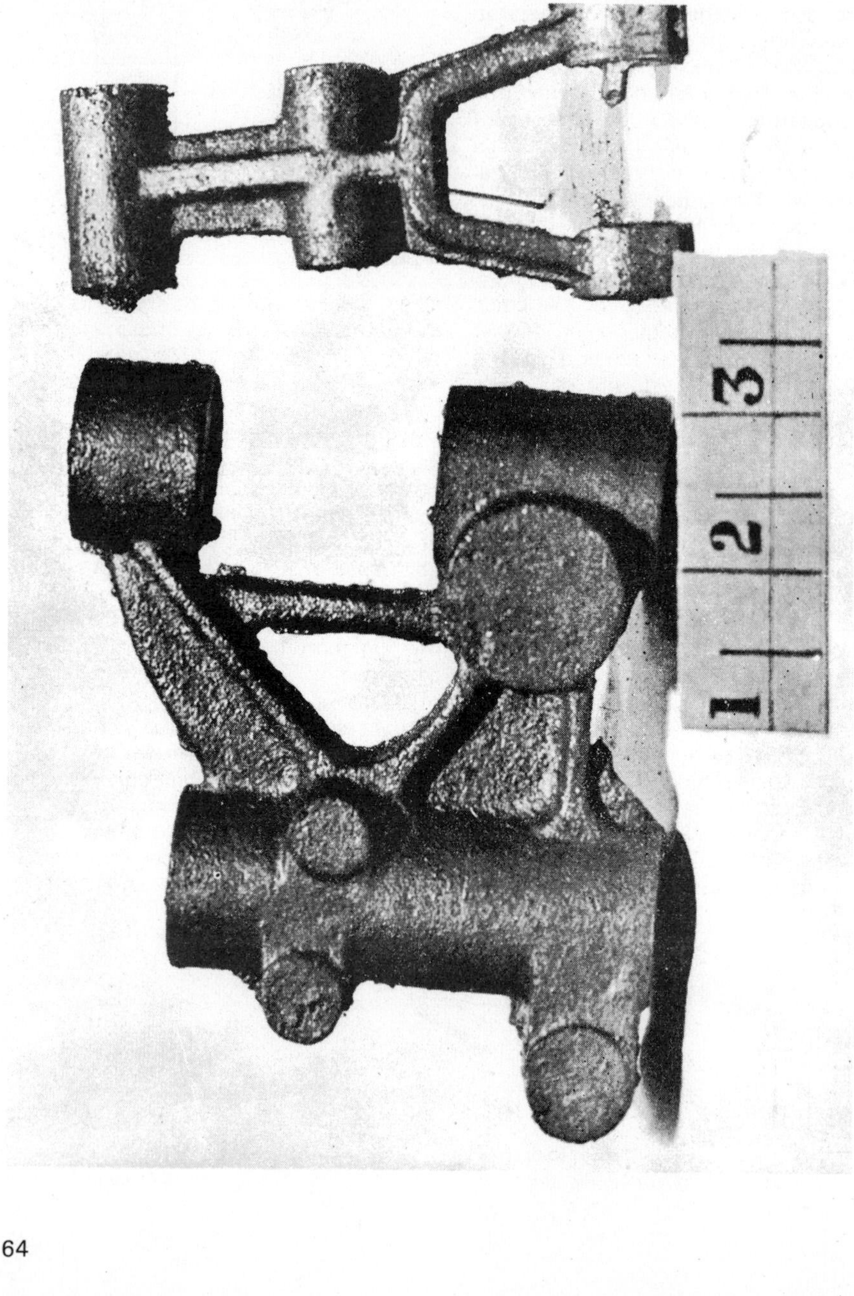

Unfettled headstock and jockey bracket castings, showing flash at the joint line of the mould.

the ingate will be related to the bulk of the casting, of course, and it is probably safer to err somewhat on the generous side. Laborious fettling is much to be preferred to a mis-run caused by a restricted flow.

The patterns are loosened by "rapping." A common form of rapping bar is illustrated in Fig. 68 and, for small patterns, may be made from quarter to

Fig. 68. Rapping bar.

half-inch diameter steel rod. It is given a light tap to drive the pointed end into the centre of the pattern and a second bar, known as a "striker," is employed as in Fig. 69. It is important when withdrawing the pattern that it should be lifted at as near as possible the point of balance, to minimise the risk of dragging the mould. A useful accessory in this direction is a "pot" hook or screw eye, which is screwed into the pattern after rapping.

Becoming accustomed to the technique, a moulder will be able to "feel" if a pattern is quite clear all over or whether in need of further rapping. With patterns having a deep hollow on the underside there is a tendency for the core of sand to adhere in the space and break away from the mould. A few sharp taps with the striker on top of the lifting hook, just at the moment the weight is taken, will usually have the desired effect and, where the contour is particularly delicate, a light rapping can be continued until the pattern is largely clear.

With both patterns removed the mould for the machine vice appears as in Fig. 70; ready, now, for dusting with plumbago before final closing. This is the stage where, if any, cores would be fitted. The mould complete, weighted ready for pouring is shown in Fig. 71. The weights in this case are a couple of "wasters" from

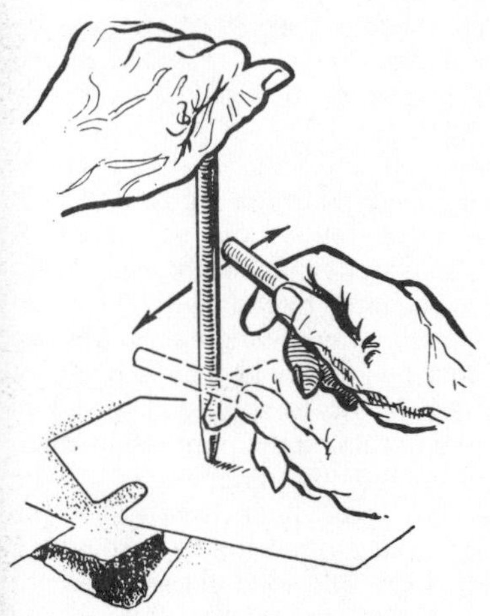

Fig. 69. Use of rapping bar and striker.

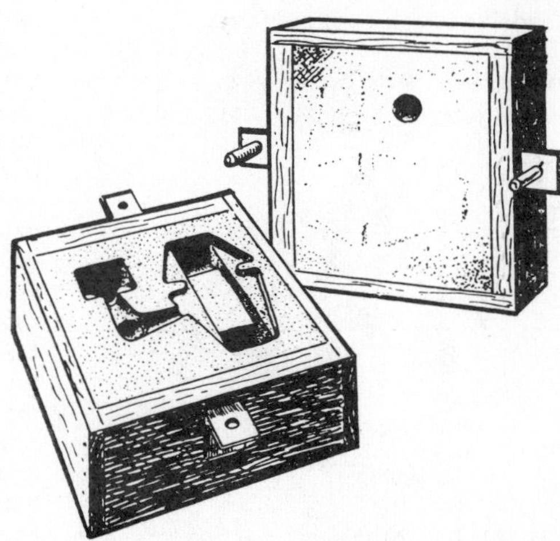

Fig. 70. Mould for machine vice.

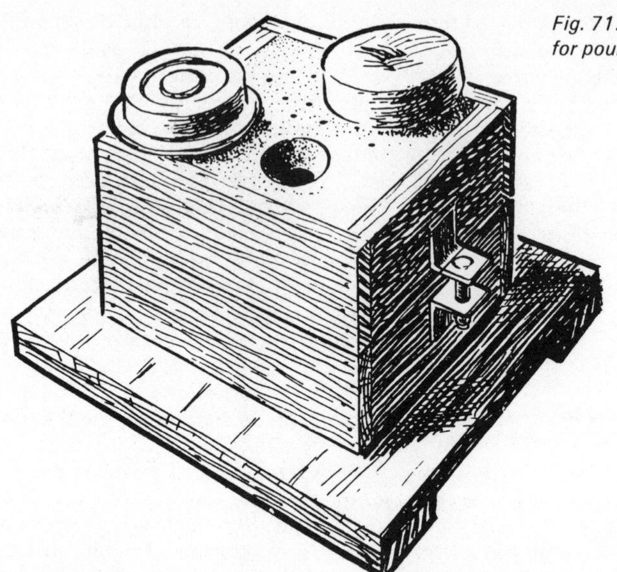

Fig. 71. Mould, weighted, ready for pouring.

earlier foundrywork. The castings for the machine vice and loose jaw, with runner attached just as they leave the mould, are shown in Fig. 72.

Oddsides

The foregoing has dealt with the moulding of patterns having one flat side. Patterns which have an all round contour very frequently present a different kind of problem. Where these are dealt with by splitting them at a convenient parting line they, too, can be treated from the point of view of ramming up the drag, as flat sided patterns. In this case the part of the pattern without pins is moulded in the bottom box and, before ramming the cope, the second half of the pattern is registered over the first. The two sections of pattern divide with the mould and each half has to be rapped out separately. A point to be watched is that the dowel pins do not become fouled with sand to prevent them withdrawing easily. Many a mould has been scrapped as a result of the pattern failing to split, pulling out of the sand and ruining the impression when the cope is lifted off for rapping.

Reference has been made, in the chapter on patternmaking, to methods of moulding solid, oddsided patterns using a roughly made template. While the arrangement will be satisfactory when

Fig. 72. Iron casting for machine vice with runner attached as it leaves the mould.

Drilling machine of $\frac{1}{4}$" capacity cast in iron. Fig. 40 shows the construction of the headstock pattern and pages 60, 62 & 64 stages in casting.

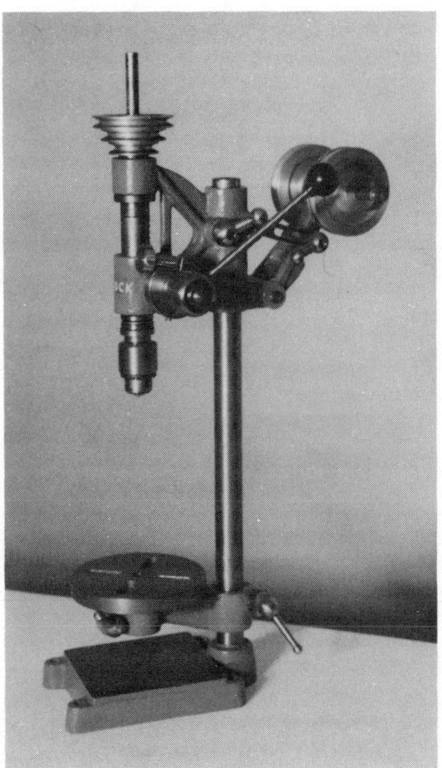

dealing with patterns having moderate relief, it is sometimes necessary to deal with deeper sections in a different manner and the word "oddside," in foundry jargon, is usually applied to the kind of specially prepared bed made to receive one half of the pattern while the first box part is rammed up. Sometimes the oddside, as a temporary measure for "one off" jobs, is made in greensand. A box part is prepared by filling with sand and levelling the top. A cavity is dug out of the sand to receive the pattern up to the parting line, where it is neatly embedded and the sand trowelled about it to make a clean joint. The whole is dusted with parting powder and the drag box, inverted over it, rammed up in the usual way. For removing the oddside, both boxes are turned over together and the upper one, emptied, is replaced for ramming the cope.

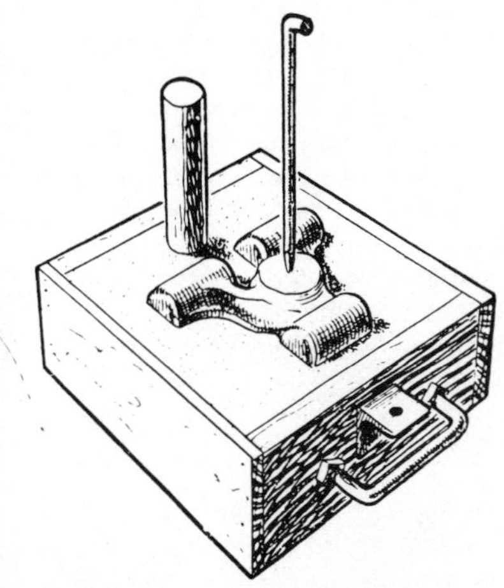

Fig. 73. Rapping bar placed in position on oddside pattern before cope is rammed.

From this stage onwards moulding proceeds as with a split pattern, except with regard to the use of the rapping bar which, in this case, is attached to the pattern before the top box is filled. Thus it protrudes from the sand at the top of the mould, allowing the pattern to be well loosened within before the box is separated. The cavity left in the cope when the rapping bar is withdrawn can remain to form the "riser."

When a more permanent form of oddside is required in a case where a fair repetition of the casting is anticipated, it

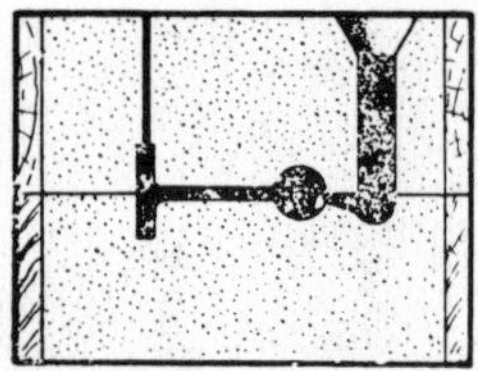

Fig. 74. Mould with riser.

can be made, very conveniently, in oil sand and baked hard. Such an oddside will, if used with care, serve as a basis for more than one mould. Of an even more permanent nature, an oddside may be made from plaster of paris, and varnished against erosion. The same may be used itself as a pattern, for *casting* a metal oddside, but a job likely to call for one so permanent is not frequently encountered in an amateur foundry.

Risers

Except with the reference to a riser in the paragraph above, the reader has been left to assume that the only connection with the internal cavity of a mould is via the runner. This, with regard to the kind of small, flat sided pattern in the example of mould making already discussed, is largely true. Where the upper part of the cavity is in line with the joint of the mould, trapped gases, which may otherwise impede the flow of the metal, will readily dissipate. It is when a pattern leaves deep pockets in the cope that attention must be given to releasing trapped gases by providing a riser. There is no need for this to be more than a small hole, made with the venting tool, which is pushed through from inside the mould. It is not necessary that metal should rise up it to indicate that the mould is full as this is usually apparent by the filling of the runner.

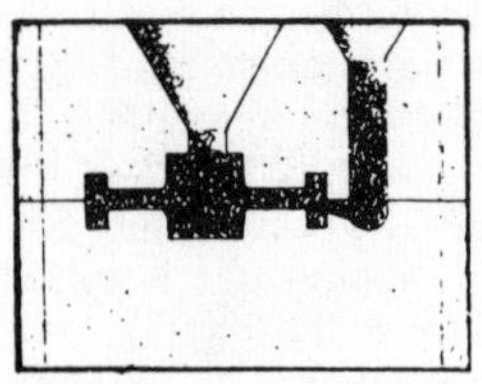

Fig. 75. Mould with feeding head to ensure soundness in the centre section.

Feeding Heads

Sometimes, however, where a pattern has a particularly heavy section likely to be difficult to run soundly, due to contraction, it is desirable to combine the duty of riser with that of "feeding head." In this case the riser, where it connects with the cavity of the mould, is made of much more substantial section, comparable with the runner, and it is opened out above to form a wide funnel, Fig. 75. When the mould is poured metal rises in the feeder where, due to its large volume, it remains liquid while the casting is in the process of solidifying. More metal can be added, by shifting the pour from the runner, and the neck of the feeder kept open by "pumping" slowly with a plunging rod.

When dealing with castings of that nature, too, it is often desirable to increase the height of the runner or sprue head, to add to the pressure of metal by gravity and, where a feeder is employed, the height of this must be increased to correspond. A reference to Fig. 76 will make this point clear. Where such a runner bush is employed the pouring cup is omitted from the mould itself but is provided in the extension. Convenient bushes can be made from greensand rammed into a metal ring and one of the flatter type of food cans, with both ends cut out, will serve admirably. A few of these can be acquired, in various sizes perhaps, at no cost, forming a very useful accessory to the moulding shop. Feeding tubes moulded from special compounds, having a "thermic" (heat producing) effect on the molten metal, are obtainable from foundry suppliers.

Moulds That Are Dried

While, basically, the same technique is employed as for greensand, there are one or two important differences in the use of oilsand, for moulding purposes outside of general core making. From the point of view of the home foundry the greatest obstacle in the way of oilsand moulds is the necessity for baking them out before use. Such moulds cannot be used damp or even allowed to air dry, as their particular value lies in the fact that, baked hard, their strength is very much greater than that of a greensand mould and they can be used for castings of a very intricate nature, in which duty the sand of an ordinary mould would probably collapse.

Where an occasional oilsand mould is required by the amateur it will be possible to make it in one of the usual wooden boxes and bake it, box and all, in the domestic oven. The temperature required, as for cores, is around the 400°F mark and, although the wood will char a little it will still be quite serviceable. A mould made in box parts each 6″ × 6″ × 3″ or thereabouts, will take between two and a half and three hours at this temperature to bake out thoroughly. A previous experiment with the oilsand to be used will indicate, by its colour, when it is quite adequately baked. Unfortunately, the baking of oilsand moulds in this way gives rise to something of an odour – partly from the wood and partly from the bonding agent – which may make the process objectionable from the domestic angle but, perhaps, careful attention to ventilating the kitchen, coupled with the choice of a convenient opportunity, may go a long way towards a solution of the problem so raised. The alternative of equipping the foundry with a small core drying oven is not to be overlooked. An oven built of brick behind the furnace and heated by the furnace flue would be very useful indeed and not at all a costly addition. Much, of course, depends upon the size and situation of the amateur foundry and the ingenuity of the foundryman. If the household cooker it must be, perhaps some consolation can be gained from the fact that it is yet within

Fig. 76. Mould with extended sprue and feeder.

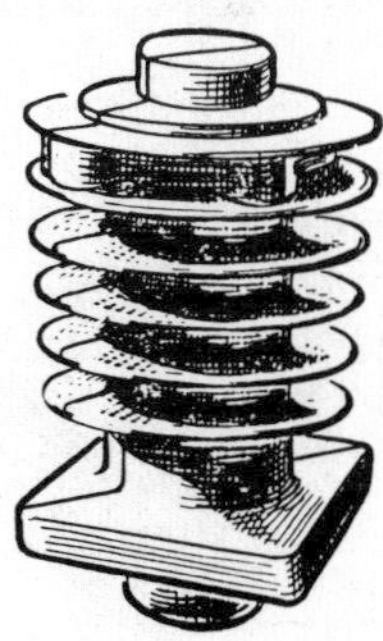

Fig. 77. Pattern of finned cylinder for moulding in oilsand.

living memory that the jobbing foundryhand took home his cores for baking!

The use of snap flasks for this kind of work has an obvious advantage over the standard variety as, in that case, only the sand of the mould is actually conveyed to the oven. The simplest way of doing this is to mould and extract the pattern, apply the blacking, fit the cores and close the mould again, ready for pouring, before removing the flask. In this way the mould is baked intact and there is no need for handling the separate parts after they have been deprived of the support of the box.

Where the parts of the mould *must* be baked separately, it is usual to fit locating pegs to the mould faces themselves and to insert metal pieces in the sand at the sides of the cope for handling. The sand after baking, however, is really quite hard and firm and there need be little danger of breaking it. Before pouring, it is advisable to moisten the sand round the mould at the joint line as a precaution against a run-away. Metal which may penetrate the joint would solidify fairly rapidly in contact with the moisture and thereby seal itself. Nevertheless, rather more care should be exercised in weighting to top of *any* mould removed from a snap flask.

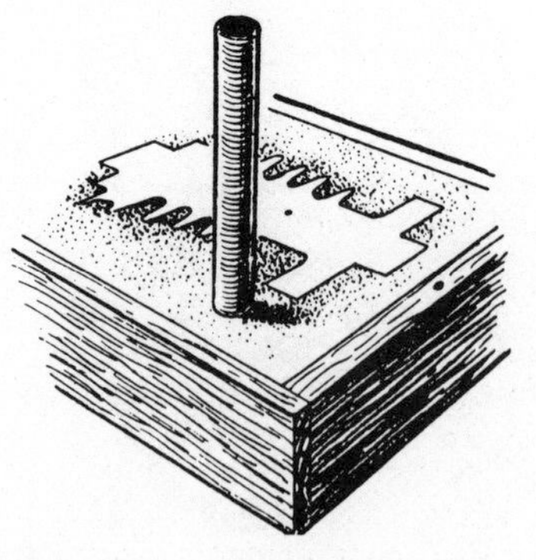

Fig. 78. Runner tube positioned close to base flange.

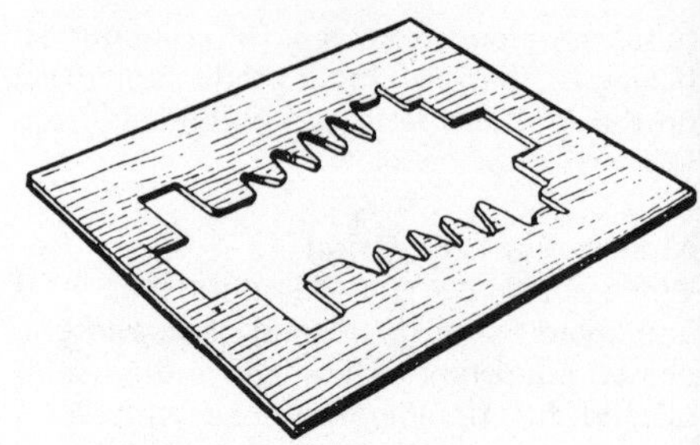

Fig. 79. Stripping plate.

Moulding

The use of a liquid parting, such as *"Separit"* is especially recommended for oilsand moulding. It is applied to the patterns, by brush or spray, before ramming and it affords an extremely clean strip. From the point of view of economy the oilsand is used only in actual contact with the pattern, the remainder of the box being filled up with the normal greensand of the foundry. The example chosen is a finned cylinder pattern of a type universally cast in a drysand mould. The pattern, of course, is split and the method of moulding described earlier will be found most effective when dealing with the fins. The impact forces the sand well down between them and a little pressure of the hand will make the sand quite firm. With the pattern covered to a depth of about half an inch with oilsand, the box can be filled in and rammed in the ordinary way. If venting has become a habit, well and good, but it is not really essential in the case of very small, dried sand moulds. The drag is turned over and it will be seen from Fig. 78 that the runner is positioned close to the base flange of the cylinder.

After ramming up the complete mould the patterns must be removed. Rapping will have to be thorough but the accent

A pair of 2-stroke cylinders cast in iron

should be in the direction of the fins rather than across them. The suggestion of tapping vertically on the lifting hook at the moment of withdrawal applies here. Another useful accessory in the direction of a clean draw is the "stripping plate," Fig. 79.

If the half of the pattern without the pins is laid on a sheet of thin plywood and a pencil line drawn closely round its outline the result, when cut out with a fretsaw, will be a stripping plate near enough for practical purposes and, as this represents the work of only a few minutes, it is a worthwhile insurance against the possibility of a difficult draw. When the pattern is being lifted, only a slight amount of sand adhering between, perhaps, two of the fins will ruin the mould. With the stripping plate held down to the sand with one hand, while the pattern is drawn with the other, risk of damage is very much reduced. Another advantage is that, with the plate firmly in position the pattern cannot be lifted at the wrong angle. The plate, reversed, serves equally for either pattern half.

Every effort should be made to withdraw the pattern cleanly at the first attempt. An untouched mould is always the strongest but, human nature being what it is, most moulders will try to effect a repair to a damaged mould rather than scrap it and start again. When it turns out

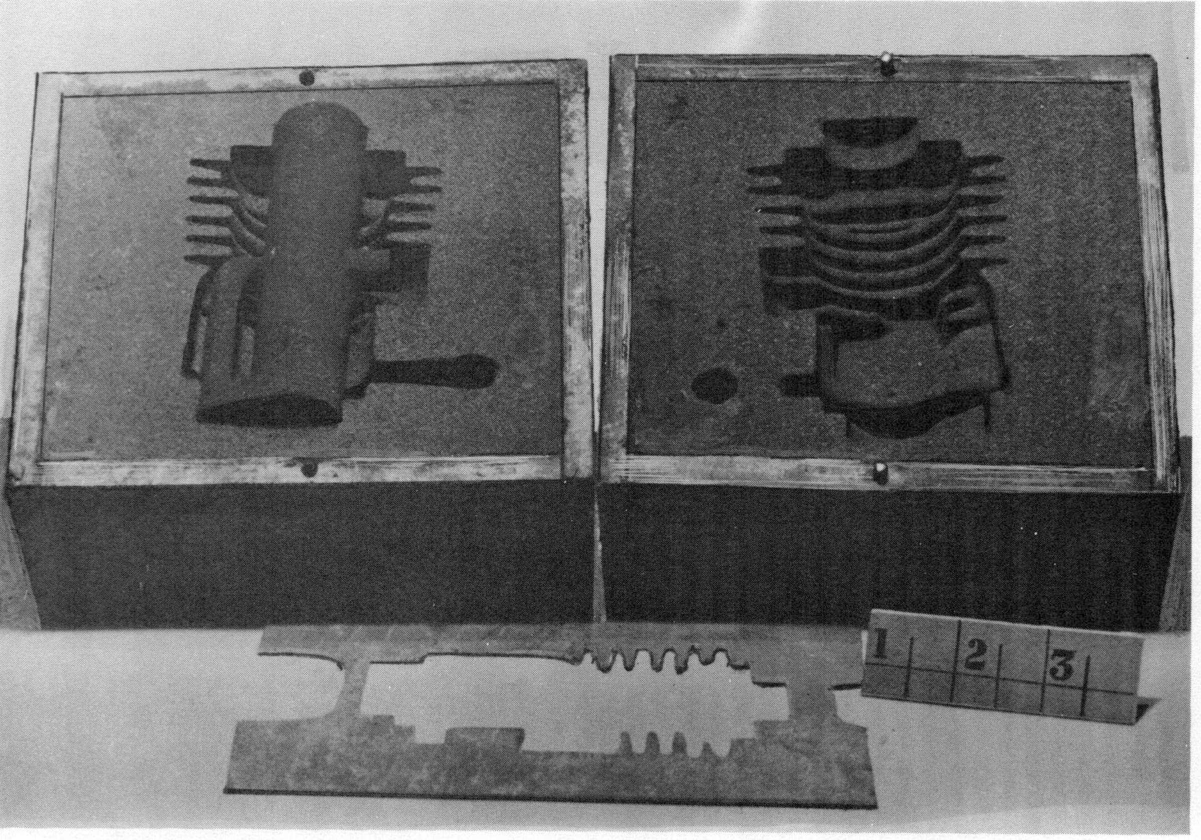

Mould & core assembly in oilsand for the casting in iron of an engine cylinder. Note stripping plate.

complete, except for a tiny fragment of fin, which has come away with the pattern, the temptation is very great. In many such cases success may be obtained in the following way. Remove the offending sand from the pattern and examine the doubtful area for any suggestion of roughness, which may have been the cause. A light touch with smooth glasspaper will probably put the matter right and a fresh application of parting medium – powder or liquid – will help towards a cleaner draw. A pinch of oilsand is then flicked into the area in an amount slightly in excess of the missing part and the pattern returned carefully to the mould. To make sure that the new sand remains in place this time, a direct rapping over the spot at the moment of lifting is again recommended. If the operation has been successful the repair will be hardly visible. A little extra moisture applied to that part will improve the "knit" and, after baking, it will be quite as strong as the rest.

The employment of a riser, suitably pierced from one of the upper parts of the cavity – the position of the bolt pads at the top of the cylinder is a likely spot – will release gases generated in the mould during pouring, before a build-up of pressure is able to cause damage.

If powdered blacking is to be used this should be applied while the mould is still green. Liquid or paste mould washes are best applied *after* baking, probably calling for a further short period in the oven. Spirit based mouldcoats evaporate fairly rapidly in the air or can be fired. From here the process is as already described.

CHAPTER 9

Melting Procedure

Of the many stages involved in the production of a casting, there is little doubt that it is in the melting of the metal and the pouring of the mould that the interest runs highest. The process is simple enough in all truth, but it is also true that it does call for some care and not a little skill in the manipulation of the furnace and the metal if the results are to be satisfactory. Whether the foundryman prefers to make all his moulds at one time and cast at another, or whether he does his moulding and casting at the same session, it is always the best policy to have his mould quite ready before commencing to melt the metal for it. It is generally acknowledged to be difficult to do two things successfully at once and, in the matter of foundrywork, the fact is more than particularly true.

Without becoming deeply involved in the science of metallurgy, calling for an enormous outlay on apparatus as well, there is little prospect of the amateur being able to regulate the composition of his metals to any fine degree. For one thing his charges will, almost certainly, be made up of scrap. Ingot metal would be expensive. But by no means should this be taken as an indication that amateur castings will be, of necessity, anything but serviceable. On the contrary, and, if one delves only a short way into history, the evidence is clear that, until comparatively recent times, foundrywork in one form or another has been carried out from the very beginnings of civilisation without the advantage of metallurgy, as a science, being even suspected. There is, in fact evidence that in the past superstition akin to witchcraft has been associated with foundrywork, to the extent of adding frogs' legs and mouses' ears to the melting pot, to coax the metal to behave as the caster wished it. With the exercise of a little care, patience and common sense perfectly sound castings, in practically any metal, can be made and the following can be regarded as an indication of some of the major points to be watched.

Iron

The first consideration is the source of supply. Iron, fortunately, is very abundant and almost any old casting can be used. In as much as most of the patterns likely to be moulded will be of a comparatively thin section, it may be as well to choose scrap of a similar nature. Usually, this will be of a composition chosen originally for its suitability to that class of work. Another consideration is, of course, that thinner scrap is more easily broken up for filling the crucible. That does not mean that the smallest possible pieces are ideal. Rather, it is sometimes an advantage to use as large pieces as can be conveniently accommodated in the pot. The reason is that, in the earlier stages of a melt, considerable oxidation takes place on the surface of each fragment and the proportion of oxide, compared with molten metal released, will be greater the smaller the pieces. Either way, however, it is desirable to charge the pot initially, with as great a weight of metal as it will contain, without cramming it so tightly that there is no room for expansion. It is more convenient to do so while the crucible is cold when, carefully filled at first, only one further replenishment may be necessary to obtain a pot nicely topped with molten iron.

One advantage which might be claimed for the use of a crucible for melting iron – compared with the normal procedure of cupola melting – is that the charge is not influenced to any great extent by the pick-up of furnace gases. Neither are there any appreciable losses, so that the iron when melted can be regarded, to all intents and purposes, as being in close similarity to its original composition. Thus, if a metal is chosen for any particular characteristics (as an example – a scrap cylinder block for re-casting as a miniature cylinder) it can be taken that its suitability will be, largely, retained. Nevertheless, some control may be desirable and *is* possible, about which more later.

It is of some importance to heat up the metal as rapidly as possible and, to this end, there is an advantage in having the furnace at close to white heat before introducing the pot. Usually it is as well to blow up a furnace, fairly full of coke, to a good heat before trying to melt anything. This charge having been partly burned down, the remaining fuel can be rammed down fairly densely at the bottom to form a solid bed on which to stand the crucible. Neglect of this point will probably result in a hole being blown right through the fuel at an early stage of the melt, with a consequent reduction of temperature instead of an increase. At this stage the top of the crucible should be just below the level of the furnace mouth and small coke should be packed well round it. With the furnace cover replaced the blast is applied.

To reduce the iron to a fully molten state can be expected to take from ten to

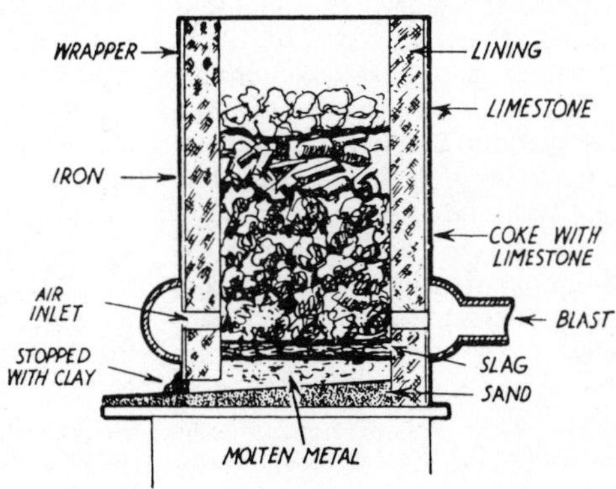

Fig. 80. Diagram of small cupola for melting iron.

twenty minutes, according to the size of the vessel used. Progress can be observed by the gradual changing in colour of the exhaust flame, from blue to yellow, when, with the blast still on, the furnace cover can be lifted slightly and the progress noted from the appearance of the pot. Although the crucible itself may be nearing white heat it should be allowed to remain, undisturbed, as long as the space between itself and its lid shows darker than the rest. It is not until the interior of the furnace appears fused into one brilliant, white mass that the iron can be assumed to have reached a temperature at which it can be poured.

If the quantity of metal is sufficient for the job well and good, but, in perhaps the majority of cases, more metal will be required than has resulted from the primary charging of the crucible. If so, cut off the draught, push back the furnace cover and *lift the crucible lid*. The fuel will also have to be replenished before proceeding and, during the interval when the pot is lifted a little while small coke is packed underneath and round it, the molten slag, which now covers the crucible, will have hardened somewhat. This would have made the lid very difficult to remove later. Fresh metal should be added with care. It is extremely dangerous to *drop* pieces of iron into the pot. Slight dampness can cause an explosion and, in any case, no one wants splashes of white hot iron flying about. Often it will be found that if a fairly large piece is placed in gently first it will tend to solidify the top of the melt and further additions can be made without risk. The less time occupied in re-charging, the smaller the loss of heat and usually it will be found that the second charge melts a good deal quicker than the first.

In full scale foundry practice the charging of a cupola takes the form of coke, iron and limestone added in layers. The limestone melts and forms a liquid slag, which fluxes away the impurities and, at the same time, floats on top of the molten iron providing a protective coating. In the crucible melting of iron, limestone is not used. A liquid slag can be obtained, however, by the use of a material such as *FerroFlux* added to the charge at the commencement of melting. This has the same protective effect, dispensing with the need for a lid and, at the same time, bringing the dross away with it when it is skimmed off prior to pouring. Such additional items, while not indispensable, go a long way towards the improvement of amateur castings while their cost, on the other hand, is negligible.

To ensure that castings of comparatively small section are easily machinable it is important to keep the silicon content of the iron fairly high. High silicon content makes for a reduced tendency to chilling, which is the enemy of the turner and the ruination of his tools. Evidence of chill in an iron casting is discernible in the fracture where the runner is broken off. Likewise, when an attempt is made to saw off the runner instead, the hacksaw blade is blunted before the slightest impression is made on the metal. With the re-use of scrap metal the silicon content is not likely to be known and, as a guard against wasted effort, through a casting being difficult or impossible to machine, it is a simple matter to make small additions to the molten metal in the crucible immediately before pouring. A material is obtainable under the name of *Ladelloy*, which is normally placed in the bottom of the pouring ladle in which metal is carried from the cupola. For making the addition to a crucible, a small quantity is dropped on top of the metal after skimming and stirred in. There is a slight thermic reaction which tends to increase the temperature of the melt. Fluidity is also improved, a condition of especial value where very thin sections are being cast.

That the exact suitability of the iron for the job on hand is difficult to gauge has been admitted but, with due regard for the conditions outlined above, it will be found that articles, crucible cast in the manner described, will be composed of a fine grained, grey iron delightful to work with in a small machine shop.

Handling Molten Metal

A type of simple crucible tongs, easily made, is shown in Fig. 81. The example illustrated was made from ⅜″ reinforcing steel, flattened at the centre to give spring to the bow and at the ends, to form the jaws for embracing the pot. Half inch flat iron would have been equally effective. The jaws are bent outwards to an angle which allows a measure of purchase to be obtained on the crucible in the furnace from a vertical position and, at the same time, permits the handling of the pot for pouring. The length should be at least two feet and, even then, a pair of gloves is an absolute necessity when dealing with iron. Added security is gained by the use of an S shaped ring or link between the arms. It is pushed down when the jaws are closed on the crucible, taking some of the strain off the caster's wrists. Fig. 82.

While such tongs can be used for crucibles of varying sizes, an ideal form of tongs as in Fig. 83 is made specifically for one size of pot. This pattern is admirable for lifting pots from the furnace but it is usual to use it in conjunction with a ring shank. Fig. 84. That with one handle is for small pots and the one with two handles is intended to be carried by two men when the weight of metal demands it. The crucible is transferred from the tongs to the shank for pouring.

Stirrers, plungers and skimming ladles should all be of good length and made from 3/16″ to ¼″ mild steel or iron wire. One end should be provided with a loop for a grip and, for a prolonged life, the business end should be coated with a

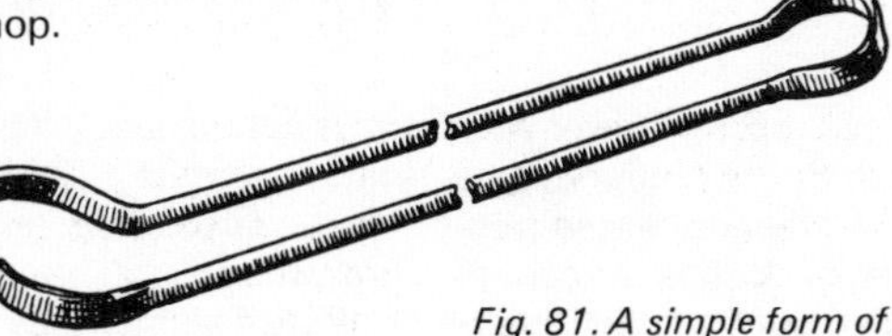

Fig. 81. A simple form of crucible tongs.

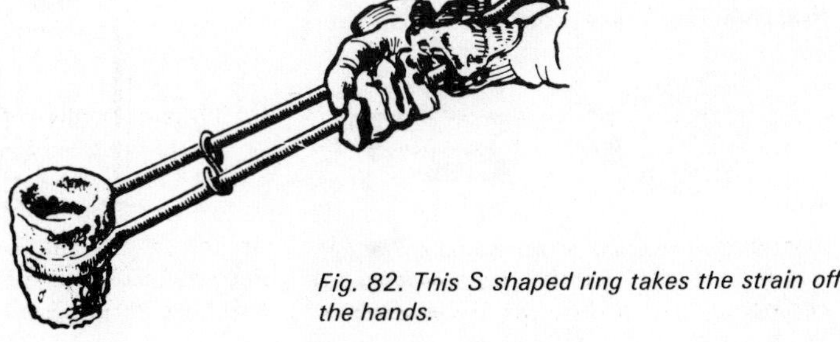

Fig. 82. This S shaped ring takes the strain off the hands.

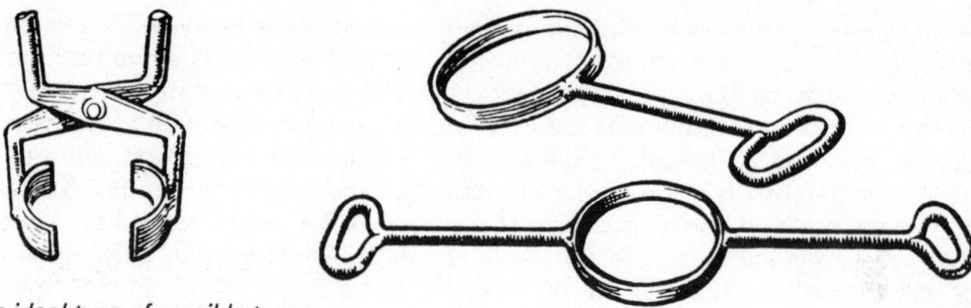

Fig. 83. An ideal type of crucible tongs.

Fig. 84. Ring shanks for pouring.

refractory wash. In Fig. 86 the stirrer A, is simply a straight rod. B, the plunger has a cup turned from mild steel or beaten from steel sheet. The ladle C has a spoon-shaped extremity also cut from mild steel sheet and riveted on.

Aluminium Alloy

The low melting point – some 1,200°F (650°C) – of aluminium and its alloys makes this material of great general utility to the amateur foundryman. It often enables him to make castings which, in other metals, if only on account of their size, would be outside his scope. At the same time it can be claimed for many alloys of aluminium in use today that their strength is equal, in many ways, to that of iron or even steel. Except where either of the other metals has a functional value, light alloy can in fact, be substituted for many of the light machine jobs typical of the model maker's workshop. Engine bedplates, plummer blocks and bearing brackets, crankcases, pulleys, handwheels to mention but a few. Often too, where duty permits, aluminium components housing rotating shafts can be used unbushed.

Pure aluminium is never used for castings. As referred to here the metal is invariably alloyed with, mainly, zinc or copper; sometimes with both and, sometimes, with small additions of nickel and other metals. The actual alloying, however, is likely to concern the small scale foundryman very little since, again, his most probable source of supply will be scrap. Very often, none the less, he will be able to control the composition of his

Fig. 85. Two sections of iron runners. The right is free from chill. The left bears evidence of chilling.

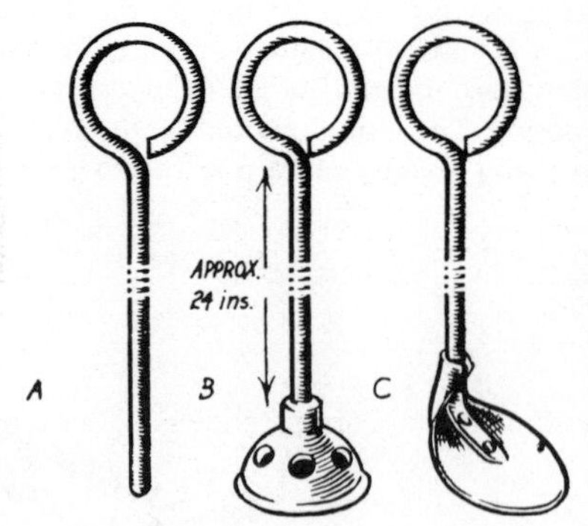

Fig. 86. Tools for stirring, plunging, and skimming the dross.

Pouring iron, photographed only by the light from the crucible. Note the iron weights to prevent the cope from lifting.

castings by a careful choice of suitable scrap. The motor car seems to provide, when broken up, perhaps the most abundant supply of aluminium and it is to the repair garage or car breaker that the amateur frequently turns. The usual light alloy components will be well known and pistons, in particular are extremely valuable for re-melting, if only on account of the fact that the smaller ones can be popped straight into the pot without breaking.

Occasionally, large pistons from lorry and bus engines will be found to be so tough as to resist all efforts to smash them up with a hammer. The answer in this particular case is to heat them up in the furnace almost to red, when they will break with ease.

Unlike iron, there is not much advantage in charging the crucible before placing it in the furnace. Where the draught to the fire is what it should be there will be no need to apply blast and it is a good policy to heat the pot to a bright red before charging it at all. It is readily accessible all the time so it is quite practicable to keep adding pieces as the metal melts. Once a pool has been formed in the bottom, added metal melts into it very rapidly.

Although too high a temperature is detrimental to aluminium, so is a prolonged period of melting. "Stewing" should be avoided at all costs and the policy of having the moulds in complete readiness is particularly important when dealing with the metal. The best all round results are obtained with a furnace that is heated *adequately* – bright red – so that the alloy melts quickly and the fresh pieces, added at frequent intervals, prevent the temperature from rising much above melting point. The warning, in Chapter 2, against carelessly dropping large pieces into the pot when replenishing, is repeated. The metal, when right for pouring, should appear, in the

An uncluttered floor and a firm stance are essentials when moving and pouring molten metals.

Above, The impression in the drag with cores in position.
Below, the impression in the cope.

The completed V-twin engine for which patterns are shown on page 46.

shadow, dull red and, in the light, silvery. Aluminium castings are often spoiled by the inclusion of foreign particles, so it is well worth the exercise of a little care in drossing off the crucible and, where only one pair of hands is engaged in the task, this is often carried out with less risk while the pot is till in the furnace.

Stirrers, skimming ladles, etc., like those already shown are equally suitable for aluminium but here, the provision of a refractory coating is desirable as protection against metal contamination. Special fluxes can be used and these are of particular value where dirty or oily scrap is being dealt with, while the type of plunger illustrated is suitable for the manipulation of certain degassing and grain refining media. Such items are useful, let it be noted, and may be acquired to advantage as a foundry grows but the basic requirements remain just a pot and a fire.

Moulds for aluminium

Molten aluminium possesses great fluidity and, therefore, the sand of the mould should be packed a little more firmly, if anything, than for iron to avoid metal penetration resulting in a rough casting skin. It is ideal for model work in that sections down to 1/16″ can be poured in certain cases. Realistic deep finning, for example, can be obtained on a miniature cylinder-head casting even in greensand, providing it is fed by a fair bulk of metal. Castings should leave the mould white and clean and runners and risers must be *sawn* off.

Fettling and finishing

Aluminium alloys are not easy to work with a file unless turpentine is used as a lubricant. For fettling and cleaning up castings generally, for either polishing or painting, good results will be obtained with a flexible disc or belt grinder. The oxide, which so readily forms on the surface of the metal, makes successful painting difficult. Finishing should be preceded by roughing or "keying" the surface followed by a coat of zinc chromate primer, obtainable from almost any good paint shop.

Scrap pistons have been put forward as a useful source of supply. The re-use of piston metal, however, does not ensure that the characteristics, such as strength, toughness and so on, will prevail in the new castings. In its original form the metal has been subjected to closely controlled heat and age-hardening treatment which is, very likely, outside the scope of the average home mechanic. Nevertheless, castings produced from such alloys will usually turn out very satisfactory. It is probably true that, while the mechanical properties are not so good neither are the conditions under which the parts are likely to be used anything like so arduous.

Cuprous Alloys

Castings in gunmetal, brass and bronze are used to a very large extent in some branches of model work, but even in others there are relatively few occasions when one or another of the cuprous alloys is *not* used in some measure. It is, indeed, an asset to the home workshop to be in a position to cast, at least, bronze or gunmetal stick for turning bushes and other small parts, because the purchase of such material, especially in sections of upwards of an inch or more, becomes exceedingly expensive. Moreover, the cast material is often harder and more durable, when used for wearing surfaces, than the drawn rods frequently supplied.

Although copper itself requires a temperature for melting somewhat in excess of iron, its alloys have a melting point in the region of 1,600°F (875°C) which is much lower. Like aluminium, brass, for example, can usually be melted in a furnace without the use of forced draught, but the temperature can only be reached with difficulty in an open fire. Sometimes blast, when available, can be used to advantage for quickly bringing up the heat but it should be carefully controlled as, here again, overheating can be deleterious. The alloyed zinc volatilises and is driven off as an intensely white vapour.

CHAPTER 10

A "Drop Bottom" Cupola

While in the main this book has dealt with the melting and manipulation of metals in a crucible-type furnace, it must not be ignored that, where a greater bulk of metal is required than can conveniently be handled in a crucible, it is a perfectly practical proposition to melt iron in a cupola of quite modest dimensions.

To acquaint the reader with the fundamental principles of cupola practice, Chapter Nine carries a brief description and a sectional drawing of what amounts to a small cupola or "cupolette" capable of supplying iron at the rate of about 15 cwt. per hour. Such a furnace would be beyond the requirements of even the most ambitious model engineer and thus there follows a brief description of a tiny, drop-bottom cupola, capable of meeting much smaller demands.

Let it be admitted at the outset that, although a perfectly workable cupola can be built to a very small size, no claim is made that it can be expected to perform as efficiently as its full-size counterpart. With the miniature furnace, continuous working may be limited to, at most, two or three "tappings." Slag tends to solidify rapidly below the tuyere (pronounced "tweer"), the hole in the surface wall through which the draught or blast is admitted, in the cupola of small diameter and thus there is a marked tendency for the iron in the well to cool – and solidify – as the blow continues. Nevertheless it is a small well indeed which contains less when full than an initial supply of fifteen to twenty pounds of iron. (One cubic foot of iron weighs 465 lb.) The melting of a similar quantity in a crucible would call for a furnace double the capacity of anything described in this book as well as introducing a considerable hazard in manipulation.

Construction

The method of building favoured by the writer embodies, once again, a five-gallon drum in the same manner as the crucible furnace illustrated in Chapter Two, Fig. 7. This, of course limits the dimensions of the cupola to those shown in Fig. 87, but it does solve the problem of a suitable wrapper.

Ramming up the lining follows the method described for the furnace referred to. Apertures for tap hole and tuyere should be chiselled out after the ends have been removed from the drum. During the ramming of the lining, wooden plugs shaped to suit can be inserted in these holes in the wrapper, so that they abut against the central former, allowing the

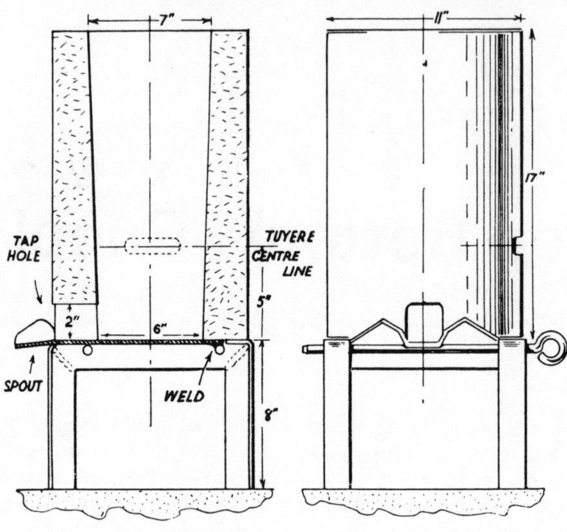

Fig. 87. Section and elevation of cupola showing construction.

packing of the clay to be carried out solidly at all points. The plugs should be made tapered in form so that they can be removed easily from the tightly packed clay and leave a clean hole. For the rest and for burning out the clay, follow the instructions already given in Chapter Two.

It will be noted that one tuyere is provided. Accepted practice in cupola design requires one tuyere aperture for each two feet of internal furnace diameter, so it will be seen that, with a diameter of little more than six inches, one vent would appear to be adequate. A second, opposing orifice could be tried experimentally, but, while the additional work entailed in building the cupola would be quite formidable, it is doubtful if any appreciable improvement in operation would be manifest. In fact, as will be shown later, it is a much greater advantage to be able to *limit* the supply of air.

In this design a "drop bottom" is provided. An alternative would be to stand the furnace barrel on a brick plinth, having sufficient height to afford clearance for the ladle under the spout. Such a cupola would require to be allowed to cool before cleaning out and de-clinkering after the melt. The hinged bottom, on the other hand, permits the rubbish to be dumped immediately the final mould has been poured and while the interior of the furnace is still red and the clinker soft. The dropping facilities are also of value in saving the furnace from complete catastrophe in the event of the metal solidifying, inadvertently, in the well.

A convenient form of construction for the base legs is $1\frac{1}{2}'' \times \frac{1}{4}''$ steel angle welded together. The $\frac{5}{8}''$ hinge bar may also be welded to the $\frac{1}{4}''$ thick steel plate forming the drop bottom. Another $\frac{5}{8}''$ bar, with a loop formed on one end, provides a

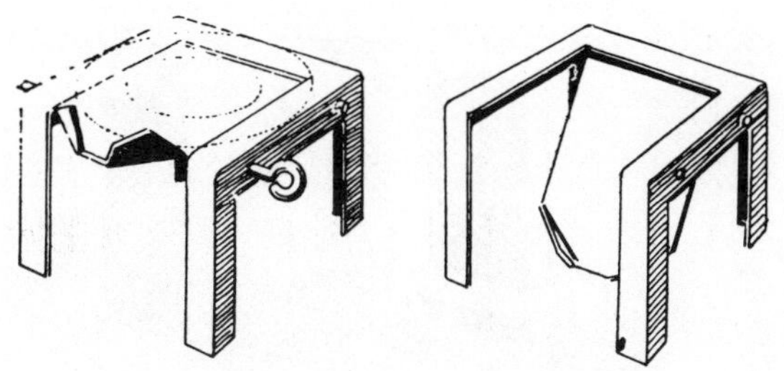

Fig. 88. The drop-bottom showing left, the working position; and right, dropped with pin removed.

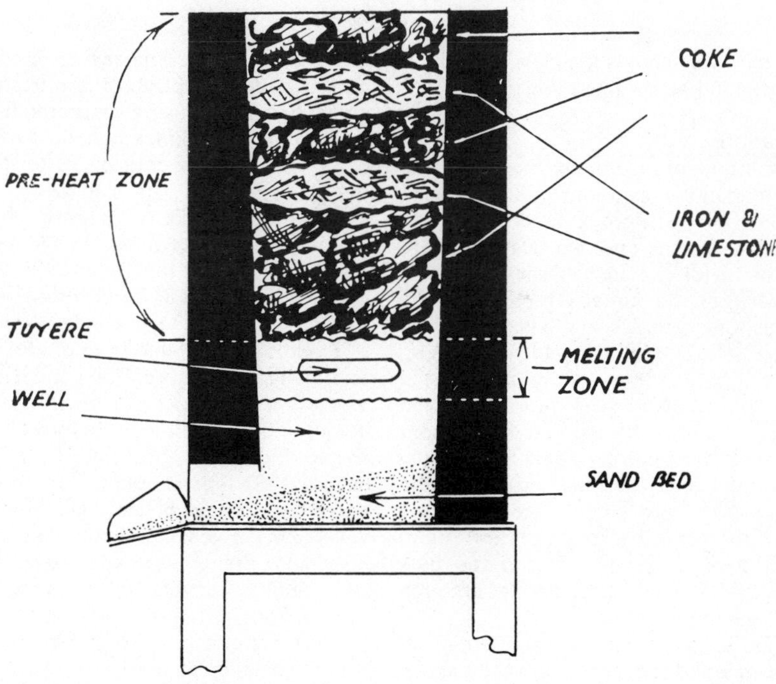

Fig. 89. Diagram of the cupola when charged for working.

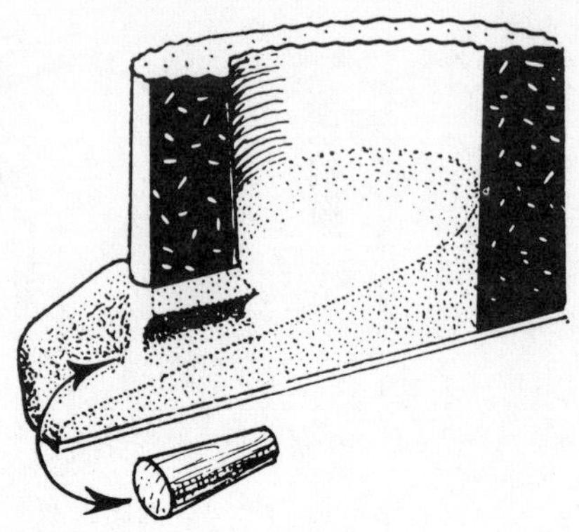

Fig. 90. How the sand bed of the well is prepared before each melt.

draw pin, which holds the plate in position while the furnace is in use, Fig. 88.

Operation

The method of operating the cupola is illustrated in the diagram Fig. 89. With the bottom in position a bed of green moulding sand is rammed over the plate to a depth of two inches, inclining from the back of the furnace to a depth of, roughly, one inch where it passes through the tap hole. The sand should be drawn up from the bed to form a gradual radius to the lining. More sand is rammed on the spout in front of the tap hole and a neat channel is moulded by hand to continue the gradual fall of the sand bed right to the lip. Fig. 90 shows this in section. Also is shown the conical wooden pattern round which sand is rammed in the tap hole itself, leaving a neat, round orifice through which the metal will run.

A fire is lighted on the sand bed, using plenty of wood and small coal or "coalite," preferably the latter. Slight draught can be supplied through the open tap hole to assist combustion and furnace coke can be added gradually until there is a good, well-lighted bed, some five or six inches above the tuyere. 20 lb of iron is charged above this in two layers separated by coke and with crushed limestone, lime or block chalk added at the rate of 30 to 40% of the coke charged. The slag which forms should be fully fluid. A "gummy" slag will be found quickly to choke up the furnace, particularly about the tuyere. Greater or lesser quantities of limestone may be used experimentally until the slag flows freely.

Blast is then applied through the tuyere and the first droplets of iron should appear at the tap hole in five or six minutes. An earlier appearance than this may be taken as an indication that the blast is too severe and steps should be taken to cut down the volume of air. Only experience will show, with each particular cupola, just how much draught is needed to obtain the best results. Conversely, of course, if the appearance of the iron is delayed appreciably beyond this, it may be taken that the blast is too weak.

Iron should be allowed to run from the tap hole until the flow becomes continuous. A further charge of coke, followed by ten pounds of iron with flux, can be

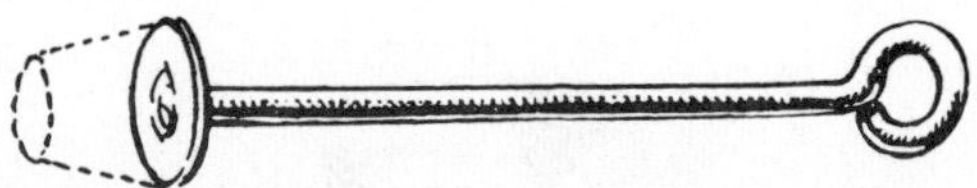

Fig. 91. A "bot" stick with a dotted indication of the shape of the "bot."

added at this stage to "keep the pot aboiling" and now the tap hole can be plugged with greensand.

With a miniature cupola it is not difficult to judge when the bulk of the initial charge is melted and some observation may be made through the tuyere if the blower tube is removed momentarily. Charging of alternate layers of coke and metal can continue until it is estimated that sufficient has been collected in the well. Here again, experience is the only practical guide.

The metal is run from the well of the cupola into a ladle for pouring the mould. For the first tap the plug of greensand is pierced with a poker and, when the ladle is full, the flow of metal is stemmed with a plug of moist clay rammed into the tap hole. This plug is known as a "bot" and is applied on the end of a metal rod shown in Fig. 91. $\frac{3}{8}$" reinforcing wire is suitable for making a "bot stick" and it should be fitted with a disc head about $1\frac{1}{2}$" diameter, which may be riveted on. The length – about twelve inches. Red clay can be used for the bot and this can be moulded on by hand, roughly conical in shape. It is useful to have two or three bot sticks ready at the commencement of each melt. A shallow bed of sand should be spread on the floor immediately under the spout to collect spilled metal and also to steady the ladle.

A crucible may be used as a ladle, but this should be preheated (to red if possible) in the exhaust from the furnace before metal is collected. But accidents are prone to occur and, if the metal should cool in the pot before it can be poured, it may be difficult, if not impossible, to remove without damage to the crucible. This could lead to an intolerable waste of good crucibles and, in my opinion, it is much better to go to some length to make or procure a metal ladle. The metal ladle itself is made of comparatively thin steel sheet and the inside is protected before use with a coating of refractory material. This lining is renewed each time before

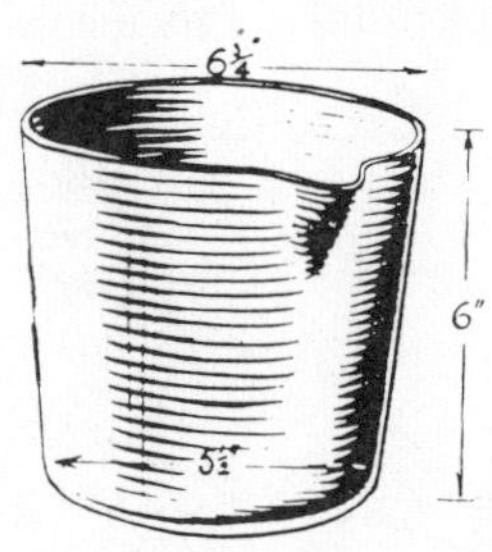

Fig. 92. A commercial, seamless ladle. Capacity after lining, approximately 20 lb.

Fig. 93. Ladle fabricated by welding from sheet steel. Approximately 12 lb. capacity.

work commences and is therefore regarded as expendable in any case.

Small, seamless metal ladles are obtainable from foundry suppliers down to a capacity of 28 lb. These, less handle or "shank," are not overly expensive, but Fig. 93 shows a small, home-made ladle welded together from 16 gauge steel. A convenient hand shank to suit and made from ½" round iron is also shown.

Normally "ganister" is used for ladle linings and this is applied in a moist state by hand and dried out completely before use. Ganister is described as a highly silicious refractory material and is usually composed of 2-3% clay, more than 95% silicon, 1% iron oxide and the remainder lime and magnesia. Like the crucible, when dealing with small quantities of iron, the ladle should also be warmed up before tapping.

Due to the effect of combustion, where the metal is melted in intimate contact with the fuel, there may be a tendency for the small cupola to produce castings with a chilled skin. Much depends on the quality of the iron with which the charge is made up. If the charge is composed entirely of scrap, as will most likely be the case in the amateur foundry, a certain amount of chilling, due to silicon loss, can be expected. The reference, therefore, already made in the preceding chapter regarding the addition to the melt of small quantities of "Silicon Ladelloy" merits attention.

Any of the devices already discussed in Chapter Two is suitable for providing the forced draught to a miniature cupola. It is, already stated, an advantage to be able to vary the volume of air to the tuyere. In this respect, assuming that an electrically

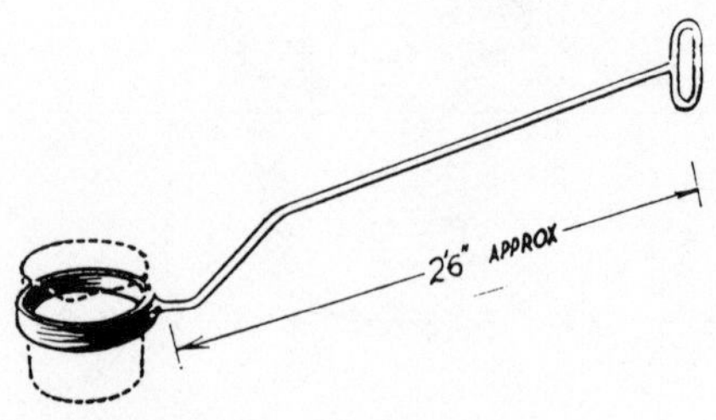

Fig. 94. Typical hand "shank."

driven blower is used and where the electric motor is of a suitable type, it is probably a more practical solution to vary the speed of the motor than to obstruct the flow of air by mechanical means. A ventilating fan controller has been used for this purpose, in conjunction with the type of blower shown in Fig. 9, with great success. The controller has a graduated panel which allows the blower speed to be adjusted to that giving, by experience, the result required. If the impeller is driven by a motor of the constant speed variety, it will be found satisfactory to interpose a butterfly-type of throttle in the delivery tube.

APPENDIX

Where Do I Get It?

Each of the manufacturers listed below will be found to be in sympathy with the modest requirements of the amateur foundryman and, often, to bestow a measure of assistance quite outside what is generally regarded as the normal run of business.

BRITISH FOUNDRY UNITS LTD.
Retort Works, Chesterfield, Derbyshire.
Moulding Boxes, Fluxes, Core Dressings, Binders, Coaldust, Sands, etc.

FORDATH LTD.
Brandon Way, West Bromwich, B70 8JL.
No. 444 Cream for core making.

F. L. HUNT & CO.
Foundry & Mill Furnishers,
48-50 Chapel Street, Salford, Manchester, M3 7AA.
Salamander Crucibles plus a wide range of useful accessories including Gloves, Plumbago, Parting Powder, Fireclay, Fluxes, Ferro Silicon, Moulding Tools and Tongs.

FOUNDRY SERVICES LTD., and JELACO LTD.
Drayton Manor, Tamworth, Staffordshire, B78 3TL.
Fluxes, Degassers for ferrous and non-ferrous alloys of all kinds as well as much useful help and advice freely given. Of particular value to schools.

MOLER PRODUCTS,
Abenbury Works, Wrexham, Clwyd, LL11 1AY.
Refractory Brick, M.P.K. (soft brick) Aggregate, Clay, Grog, etc.

Index